COGNITIVE NEUROSCIENCE

Aims and Scope of *Cognitive Neuroscience*

Cognitive Neuroscience aims to publish high quality empirical and theoretical papers on any topic in the field of cognitive neuroscience including, but not limited to: perception, attention, memory, language, action, decision-making, emotions, and social cognition. We welcome papers that link together scientific theory with current clinical applications and future clinical translations.

The journal fills an important niche in the market by publishing shorter papers (up to 4,000 words) with fast reviewing and rapid online publication (within 6 weeks of acceptance). In addition, longer discussion papers evaluate current hot topics in the field with critical analysis in the form of peer commentary (8,000 word reviews, 800 word commentaries).

Cognitive Neuroscience is not restrictive with regards to the methodologies that are considered suitable for inclusion. For example, electrophysiology, haemodynamic brain responses, and behavioural measures (e.g., reaction times) are all related to brain function and cognition, even if some are more direct measures of neural activity than others. The methodology only needs to be appropriate to the theoretical questions addressed. There are no page charges and we are able to offer free color printing where color is necessary.

First published 2011 by Psychology Press

Published 2018 by Routledge
2 Park Square, Milton Park, Abingdon, Oxon OX14 4RN
52 Vanderbilt Avenue, New York, NY 10017

First issued in hardback 2018

Routledge is an imprint of the Taylor & Francis Group, an informa business

ISBN 13: 978-1-138-41156-2 (hbk)
ISBN 13: 978-1-84872-751-9 (pbk)

Cognitive Neuroscience is included in the following abstracting, citation, and information retrieval systems:
Institut de l'Information Scientifique et Technique (INIST), Journal Citation Reports/Science Edition. Neuroscience Citation Index. PASCAL database. Pubmed. Science Citation Index Expanded. SCOPUS.

COGNITIVE NEUROSCIENCE, 2011, 2 (3–4), 135–137

Introduction

Cognitive neuroscience of bodily representations: Psychological processes and neural mechanisms

Stephen R. Jackson[1,2], Laurel J. Buxbaum[3], and H. Branch Coslett[4]

[1]School of Psychology, University of Nottingham, Nottingham, UK
[2]Department of Brain and Cognitive Engineering, Korea University, Seoul, South Korea
[3]Moss Rehabilitation Research Institute, Elkins Park, PA, USA
[4]Neurology, Department of Psychology, University of Pennsylvania, Philadelphia, PA, USA

The past decade has seen increasing interest within the cognitive neuroscience community in understanding the psychological processes involved in representing the body, and in learning how these processes may be implemented within the brain. This special issue of *Cognitive Neuroscience* presents six new empirical papers that contribute to this rapidly developing literature, together with two theoretical discussion papers that are accompanied by peer commentaries.

Recent theorizing on the nature of body representations has primarily focused on a small number of partially overlapping themes, several of which are revisited in this special issue of *Cognitive Neuroscience*.

One recurring theme within the body representation literature rests upon the notion that there appear to be multiple mental representations of the body in the brain, and has therefore been concerned with understanding the nature, function, and anatomy of each of these forms of representation. Such representations include classic distinctions, such as that drawn between "body image" and "body schema" (Head & Holmes, 1911/1912), as well as more recent conceptual entities, such as the "extrastriate body" (EBA) and "fusiform body" (FBA) areas, that have arisen as a consequence of the increasing use of functional brain imaging techniques to investigate how the brain represents the body. In this special issue, Downing and Peelen discuss the putative functions of the "extrastriate body" and "fusiform body" areas, particularly the proposal that, based upon the visual appearance of bodies, these areas code explicit representations of identity, emotion, body movements, or goal-directed actions. They argue that the current evidence is not consistent with the proposal that either region performs either of these higher-level functions, and that the EBA and FBA instead consist of cognitively unelaborated perceptual representations.

In their empirical paper, Mohamed and colleagues tackle a related question by investigating, using event-related electroencephalography, the effects that attentional load and stimulus inversion manipulations have on the processing of face and body stimuli. They conclude that while some event-related electrical potentials (e.g., N170) may reflect mandatory, category-specific, encoding in body- or face-sensitive cortical areas, others (e.g., late occipito-temporal negative components [LNC]) do not.

Lesion studies in monkeys (Rushworth, Johansen-Berg, & Young, 1998; Rushworth, Nixon, & Passingham, 1997) and in humans (Wolpert, Goodbody, & Husain, 1998), along with recent functional brain imaging studies in humans (Parkinson,

Correspondence should be addressed to: Stephen R. Jackson, School of Psychology, University of Nottingham, Nottingham NG7 2RD, UK. E-mail: stephen.jackson@nottingham.ac.uk

http://dx.doi.org/10.1080/17588928.2011.630723

Condon, & Jackson, 2010; Pellijeff, Bonilha, Morgan, McKenzie, & Jackson, 2006), suggest that the posterior parietal cortex may play a particularly important role in maintaining an accurate and up-to-date representation of the current postural state of the body (the body schema). Furthermore, it is suggested that this body schema mechanism plays a key role in the mental transformation of bodily representations during mental imagery. In their fascinating paper, Hartmann, Falconer, and Mast investigate whether instructing healthy individuals to imagine themselves as being paralyzed affects their mental transformations of imitable and non-imitable body postures. They report that imagined paralysis selectively impaired mental transformation of imitable body postures but not non-imitable postures.

To a large extent, and as illustrated in the papers outlined above, work examining the functions of the EBA and FBA consider situations in which faces, bodies, and their parts are viewed as *objects*. However, a second important and longstanding theme within the body representation literature has been to understand the psychological processes and neural mechanisms that underlie our *subjective* sense of body ownership: Particularly the distinction between the bodily "self" and the external "other", and the processes and mechanisms through which external objects may be temporarily incorporated into the bodily "self" (i.e., the mechanisms of "embodiment"). One particularly important and popular technique that has been widely used to study phenomena associated with embodiment has been variants of the so-called "rubber hand illusion" (RHI), during which, following synchronous tactile stimulation (stroking) of the unseen real hand and the seen rubber hand, the participant comes to perceive the rubber hand as feeling as if it were part of their body. Understanding of this important issue is further elaborated upon by several of the empirical papers contained in the current special issue.

Holle and colleagues investigate the necessary and sufficient conditions required to experience the RHI by manipulating factors predicted to strongly influence the process of embodiment: Postural congruence of the rubber limb with the body and the anatomical congruence of the rubber limb. Their results confirm the importance of these factors for establishing a robust RHI.

In the first of two papers that appear in this special issue, Preston and Newport examine the related process of "disembodiment" of the limb. Using a novel virtual reality device, these authors investigate how the embodiment of a fake limb is affected by the apparent encroachment onto the fake limb by an apparently solid object. Their results again highlight important limitations in the conditions under which embodiment is likely to occur.

Zopf, Harris, and Williams report the results of an investigation into the role that body ownership cues have in our ability to detect and to discriminate somatosensory stimuli. Their findings indicate that body ownership cues induce reliable differences in our perception of somatosensory stimulation.

In their second paper in this issue, Newport and Preston ask an intriguing question, namely, What happens to the real limb during the embodiment of a fake limb? While this question has been raised before in the context of RHI experiments, it is difficult to address this issue satisfactorily in standard RHI experiments because, as Newport and Preston correctly point out, typical RHI studies invariably induce a strong spatial mismatch between the viewed (fake) limb and the unseen (but proprioceptively sensed) real limb. In a clever experiment, Newport and Preston overcome this limitation using a novel VR device, and demonstrate that, in circumstances in which visual-proprioceptive mismatch is minimal, increased perceived "ownership" of a fake limb is accompanied by increased disembodiment of the individual's real limb.

A further major theme in the body representation literature is concerned with the awareness of bodily sensations and their relationship to human action. This issue is re-visited in the discussion paper by Jackson and colleagues, which focuses on the functional anatomy of the bodily sensations that we experience as urges for action. Jackson and colleagues discuss the defining characteristics of urges, with particular reference to pathological cases, such as the urge-to-tic in Tourette syndrome, and review evidence in support of the proposal that there is considerable overlap between the functional anatomy of urges associated with everyday behaviors such as swallowing, yawning, and micturition, and those urges associated with the generation of tics in Tourette syndrome.

This collection of empirical and theoretical discussion papers and their commentaries provides a timely review of the current state-of-the-art in several of the most important topic areas of the body representation field.

REFERENCES

Head, H., & Holmes, G. (1911/1912). Sensory disturbances from cerebral lesion. *Brain*, *34*, 102–254.

Parkinson, A., Condon, L., & Jackson, S. R. (2010). Parietal cortex coding of limb posture: In search of the body schema. *Neuropsychologia*, *48*, 3228–3234.

Pellijeff, A., Bonilha, L., Morgan, P. S., McKenzie, K., & Jackson, S. R. (2006). Parietal updating of limb posture: An event-related fMRI study. *Neuropsychologia*, *44*, 2685–2690.

Rushworth, M. F. S., Johansen-Berg, H., & Young, S. A. (1998). Parietal cortex and spatial-postural transformation during arm movements. *Journal of Neurophysiology, 79,* 478–482.

Rushworth, M. F. S., Nixon, P. D., & Passingham, R. E. (1997). Parietal cortex and movement 1. Movement selection and reaching. *Experimental Brain Research, 117,* 292–310.

Wolpert, D. M., Goodbody, S. J., & Husain, M. (1998). Maintaining internal representations: The role of the human superior parietal lobe. *Nature Neuroscience, 1,* 529–533.

Combined effects of attention and inversion on event-related potentials to human bodies and faces

Tarik N. Mohamed[1,2], Markus F. Neumann[1], and Stefan R. Schweinberger[1,3]

[1]Department of General Psychology, Friedrich Schiller University of Jena, Jena, Germany
[2]Department of Psychology, Sohag University, Sohag, Egypt
[3]DFG Research Unit Person Perception, University of Jena, Jena, Germany

We investigated effects of attentional load and inversion on event-related potentials to body or face distractors. Participants performed demanding (high load) or less demanding (low load) unrelated letter-search tasks. Bodies and faces were intact (Experiment 1) or without heads or eyes (Experiment 2). We measured prominent P100, N170, and late occipito-temporal negative (LNC) components. N170 to bodies had smaller and more anterior maxima than faces. N170 to intact bodies and faces was increased by inversion, relatively independently of load. Inversion effects were dramatically reduced for headless bodies, and even reversed for eyeless faces. Load effects were most prominent in LNC, with enhanced negativity under low load. We suggest that N170 reflects mandatory, category-specific initial distractor encoding in body- or face-sensitive cortical areas, a process which may depend on interactive encoding of hierarchical cues (bodies, heads, eyes). By contrast, LNC mainly reflects residual capacity allocated to extended processing of task-irrelevant distractors.

Keywords: Body; Face; Event-related potential (ERP); N170; Attention.

Human bodies and faces have been considered to recruit specialized visual processing mechanisms. Specifically, while objects are thought to be processed in a part-based manner (Biederman, 1987; Tanaka & Farah, 1993), faces may undergo configural or holistic processing (Maurer, Le Grand, & Mondloch, 2002). The well-known "face-inversion effect" (FIE), which refers to a disproportionate impairment in identifying inverted vs. upright images of faces compared to non-face objects (Yin, 1969), has been taken to support this idea.

While an FIE can be seen in increased response times (RTs) and error rates, inversion also affects the N170 event-related potential (ERP), a face-sensitive response at occipito-temporal sites that is absent or smaller for many other stimulus classes (Bentin, Allison, Puce, Perez, & McCarthy, 1996). Specifically, N170 is increased and delayed by inversion (Rossion et al., 1999, 2000). The reason for this N170 inversion effect is a matter of current debate (Sadeh & Yovel, 2010). Explanations include stronger activation of face-selective neuron populations (Rossion et al., 1999), or contributions from additional mechanisms (Sadeh & Yovel, 2010), possibly involving disinhibition of eye-selective neurons (Itier & Batty, 2009).

However, faces are not the sole category for which inversion effects are observed. Recent research uncovered similar effects for human bodies (Minnebusch, Suchan, & Daum, 2009; Reed, Stone, Bozova, & Tanaka, 2003; Stekelenburg & de Gelder, 2004). Some researchers considered this as evidence for a role of configural information for body perception, paralleling similar proposals for face perception

Correspondence should be addressed to: Tarik N. Mohamed or Stefan R. Schweinberger, Department of General Psychology, Friedrich Schiller University of Jena, Am Steiger 3, Haus 1, 07743 Jena, Germany. E-mails: tarik.mohamed@uni-jena.de, stefan.schweinberger@uni-jena.de

We thank Dr. Denise Soria Bauser for providing images of human bodies, and Kathrin Rauscher for helping with data collection. T.N.M.'s research is supported by a PhD grant by the University of Hellwan, Egypt; M.F.N.'s research is supported by a start-up grant by the University of Jena, and S.R.S.'s is supported by the DFG Research Unit Person Perception.

http://dx.doi.org/10.1080/17588928.2011.597848

(Minnebusch et al., 2009; but see Troje & Westhoff, 2006). Differences in face and body perception might be mediated by differential contributions of face- and body-sensitive regions in visual cortex: These include the fusiform face area (FFA) and occipital face area (OFA), as well as the fusiform body area (FBA) and extrastriate body area (EBA) (Peelen & Downing, 2007). Bodies and faces have been claimed to capture selective attention (Downing, Bray, Rogers, & Childs, 2004; Ro, Friggel, & Lavie, 2007). We recently investigated ERP priming effects of faces, body parts (hands), and houses under conditions of restricted attention to these stimuli (Neumann, Mohamed, & Schweinberger, 2011). Specifically, we applied different levels of perceptual load (Lavie, 2005) to manipulate attention to S1 primes. A subsequent S2 probe was either a repetition or non-repetition of the task-irrelevant S1. Face repetitions increased an occipito-temporal N250r irrespective of load, suggesting preserved face encoding. By contrast, ERP repetition effects for body parts were absent, and were not different from those for houses. Overall, while several studies consistently suggest preferential processing of faces under limited resources (e.g., Jenkins, Lavie, & Driver, 2005; Langton, Law, Burton, & Schweinberger, 2008), such evidence appears less consistent for photorealistic bodies or body parts.

In the present Experiment 1, we investigated combined effects of attention and inversion on ERPs to irrelevant bodies or faces, presented upright or inverted under high vs. low load. Specifically, we examined whether attentional capacity would be required for obtaining inversion effects to body vs. face stimuli. In view of the discussion about possible differential contributions to inversion effects of prominent features vs. holistic information, Experiment 2 manipulated attention and inversion for headless bodies and eyeless faces. In Experiment 1, considering the findings of preserved face encoding under high and low load (Neumann et al., 2011), we expected a usual N170 inversion effect, irrespective of load. To the extent that bodies engage similar preferential encoding, comparable load-independent inversion effects would be expected.

In Experiment 2, considering previous findings (Itier, Alain, Sedore, & McIntosh, 2007), we expected face-inversion effects to be eliminated for eyeless faces. Similarly, body inversion effects might be eliminated (Yovel et al., 2010) or even reversed (Minnebusch et al., 2009) for headless bodies. For conveying social attention, eyes, heads, and bodies may form a hierarchy of cues, in that order of prioritization (Nummenmaa & Calder, 2009; Perrett, Hietanen, Oram, & Benson, 1992). Moreover, very recent research suggests head posture to be an important determinant of body-inversion effects (Yovel, Pelc, & Lubetzky, 2010). Experiment 2 therefore investigated inversion effects to eyeless faces and headless bodies. By removing prominent features from the respective cue hierarchy, we sought to gain further information about the relative contribution to inversion effects of eyes (for faces) or heads (for bodies). Specifically, to the degree that prominent feature removal would compromise preferential encoding, inversion effects in Experiment 2 may be modulated by attentional load. Consequently, we manipulated attention as in Experiment 1.

EXPERIMENT 1

Materials and methods

Participants

Twenty four students (8 male) from the University of Jena contributed data (mean age = 22.7 years, $SD = 2.2$). All were right-handed, gave written informed consent, and reported normal or corrected-to-normal visual acuity.

Stimuli

Sixteen images of faces (8 female) were obtained from the CAL/PAL database (Minear & Park, 2004) and resized to 352 × 520 pixels, 7.8° × 11.5° visual angle (VA), on a 19-inch CRT monitor. Additionally, 16 frontal images of bodies (8 female), wearing neutral black T-shirts, were selected from a recent study (Minnebusch, Keune, Suchan, & Daum, 2010). Bodies were resized to 250 × 700 pixels (5.6° × 15.3° VA), and internal facial features were masked (Minnebusch et al., 2009). Images were converted to grayscale and placed onto gray background.

Inverted stimuli were generated by 180° rotation. Letter strings consisted of six red upper-case letters (Arial 26 bold), comprising target "X" or "N" and non-target "H", "K", "W", "M", and "Z" letters. Low load letter strings consisted of six identical target letters only, whereas high load letter strings consisted of one target letter (50% each) and the five non-target letters, in random order (e.g., "HKNWMZ"). High- and low-load displays occurred equiprobably and in randomized order. Letter strings were superimposed on face or body images (Figure 1).

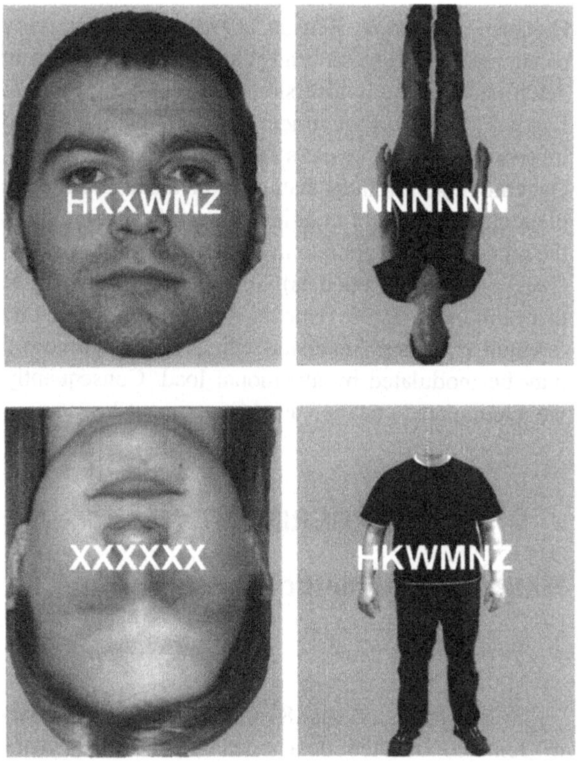

Figure 1. Example stimuli. Participants performed two-alternate choice responses ("X" or "N") to high or low load letter strings superimposed on upright or inverted distractor faces (examples in left column) or bodies (right column). Distractors could be intact (top row; Experiment 1) or with removed prominent features (eyes or heads, respectively, bottom row; Experiment 2). Note that letter strings are shown not to scale and for illustration purposes only. In the actual stimuli, letters were evenly spaced horizontally, and were presented in red color. The faces on the left of the figure were sourced from the CAL/PAL database (available at: http://vitallongevity.utdallas.edu/stimuli/facedb.html); the bodies on the right of the figure were part of the set used in Minnebusch et al. (2010) (courtesy of Dr Denise Soria Bauser, née Minnebusch).

Procedure

Participants were seated in a dimly lit, electrically shielded, and sound-attenuated cabin (400-A-CT-Special, Industrial Acoustics, Niederkrüchten, Germany). Viewing distance was 90 cm (chin rest). During each experimental trial, an initial black fixation circle was replaced after 500 ms by the blended display for 200 ms. A subsequent exclamation mark (1800 ms) allowed for responses until 2000 ms after stimulus onset. Participants made choice responses to target letters by pressing "6" and "4" buttons on a standard keyboard (numerical block), using both index fingers. Key assignment was counterbalanced between participants. Speed and accuracy were emphasized.

A 2 × 2 × 2 design included the factors Load (high vs. low), Distractor Type (face vs. body), and Inversion (upright vs. inverted). In total, 512 trials (64 per condition) were presented in randomized order, with breaks after every 128 trials.

EEG recording and processing

With a 32-channel Biosemi Active II system (Biosemi, Amsterdam, Netherlands), EEG was recorded at Fz, Cz, Pz, Iz, Fp1, Fp2, F3, F4, C3, C4, P3, P4, O1, O2, F7, F8, T7, T8, P7, P8, FT9, FT10, P9, P10, PO9, PO10, I1, I2, F9, F10, TP9, and TP10, and sampled at 1024 Hz (DC to 240 Hz), using a combined reference/ground circuit (CMS/DRL) near Cz as initial reference. Ocular contributions to the EEG were corrected using an algorithm implemented in BESA 5.1.8.10 (MEGIS Software GmbH, Graefelfing, Germany). Trials with amplitudes exceeding 100 μV or a difference between consecutive data points exceeding 75 μV were rejected. ERP epochs were quantified for 1400 ms (200-ms prestimulus baseline), recalculated to average reference, and low-pass filtered at 30 Hz (24 dB/oct, zero phase-shift). Huynh–Feldt corrections for heterogeneity of covariances were performed where appropriate.

Results and discussion

Behavioral results

Analyses of variance (ANOVAs) with repeated measures on Load, Inversion, and Distractor Type revealed that correct RTs were increased under high ($M = 777$ ms) compared to low load ($M = 543$ ms), $F(1, 23) = 190.33, p < .001$ (Figure 2). Also, error rates were increased under high ($M = 0.769$) vs. low load ($M = 0.963$), $F(1, 23) = 136.39$, $p < .001$. No other main effects or interactions were observed, all $ps > .05$.

Event-related potentials

P100. Visual inspection of the grand mean waveforms indicated the largest P100 amplitudes around 90 ms at O1 and O2. Accordingly, mean amplitudes were determined between 70 and 110 ms at these sites. ANOVAs on Electrode, Load, Inversion, and Distractor Type revealed a main effect of Inversion, $F(1, 23) = 47.17$, $p < .001$, with larger amplitudes for inverted than upright ($Mdiff = 1.35$ μV) images (Figure 3). No other main effects or interactions reached significance, all $ps > .05$.

N170. N170 was largest at occipito-temporal electrodes, though variable with respect to the precise

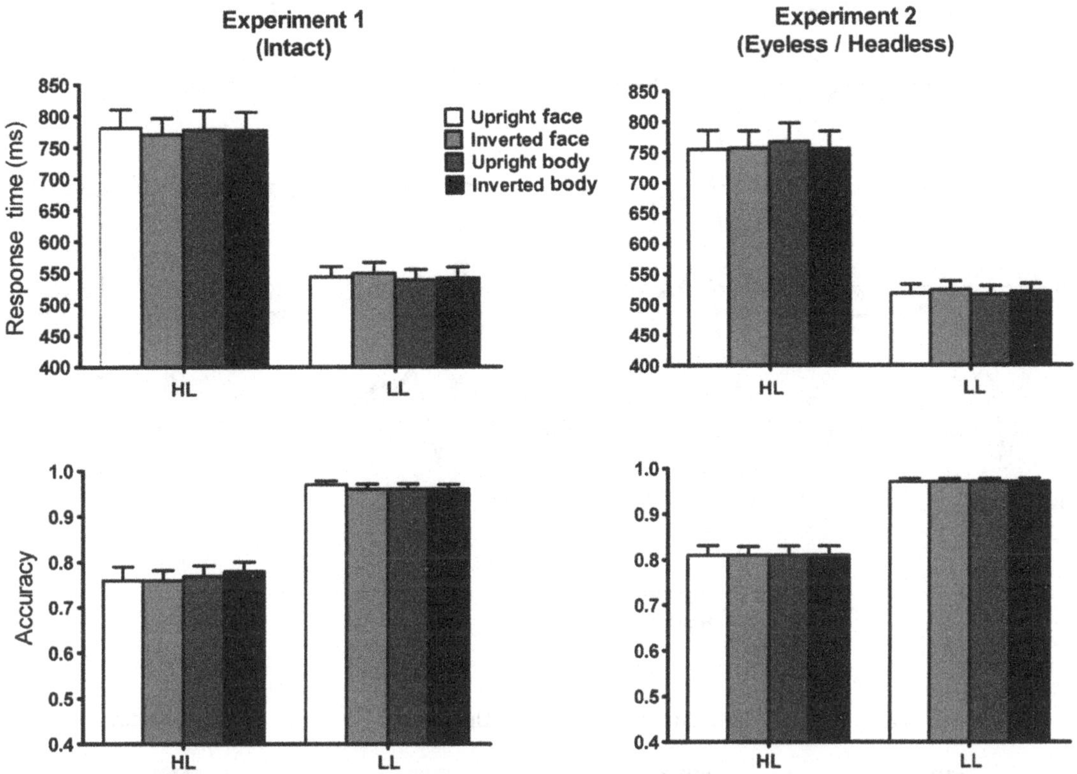

Figure 2. Mean correct response time and accuracy (proportion correct) to target letter strings in Experiment 1 (left) and Experiment 2 (right). Note the prominent main effect of load in both experiments, independent of distractor type and orientation. HL = high load; LL = low load.

spatial maximum, both between inversion and distractor type conditions, and between participants. Accordingly, we took mean amplitudes at P8, P10, PO10, P7, P9, and PO9 in the interval 120–160 ms.

The ANOVA on Hemisphere, Site, Load, Inversion, and Distractor Type revealed a main effect of Hemisphere, $F(1, 23) = 5.60, p < .05$, with larger N170 over the right ($M = -4.88$ μV) than left hemisphere ($M = -3.34$ μV). Moreover, N170 was larger for faces than bodies ($M = -5.33$ vs. -2.89 μV), $F(1, 23) = 98.35$, $p < .001$, and for inverted than upright images (*Mdiff* = 0.89 μV), $F(1, 23) = 24.13, p < .001$ (Figure 3).

The two-way interaction between Site and Distractor Type was significant, $F(2, 46) = 27.18, p < .001$. While faces elicited an N170 with a clear posterior maximum, the N170 to bodies was more evenly distributed. N170 was much larger to faces than bodies at PO9/PO10 ($M = -6.84$ μV vs. -3.48 μV) and P9/P10 ($M = -5.39$ μV vs. -2.15 μV), while this difference was reduced, though still reliable ($p < .01$), at sites P7/P8 ($M = -3.75$ μV vs. -3.05 μV).

Additionally, we found three-way interactions of Site, Inversion, and Distractor Type, $F(2, 46) = 7.46, p < .01$; Site, Load, and Inversion $F(2, 46) = 6.57, p < .01$; and a significant interaction involving Load, Inversion, and Distractor Type, $F(1, 23) = 7.29, p < .05$. Sub-sequent two-way ANOVAs on Load and Inversion were performed separately for faces and bodies. For faces, the main effect of Inversion, $F(1, 23) = 9.89, p < .01$, with larger amplitudes for inverted than upright orientation (*Mdiff* = 0.77 μV), and the interaction of Load by Inversion, $F(1, 23) = 4.37, p < .05$, were significant. Pair-wise t-test revealed a Load effect for upright faces $t(23) = 2.38, p < .05$, with larger N170 under low than high load (*Mdiff* = 0.30 μV), but no Load effect for inverted faces, $t(23) = -0.58$, $p > .50$. Moreover, while Inversion effects were larger under high (*Mdiff* = 0.97 μV) than low load (*Mdiff* = 0.57 μV), they were reliable in both load conditions when tested separately, $t(23) = 3.66, p < .001$, and $t(23) = 2.19$, $p < .05$, respectively. For bodies, the main effect of Inversion, $F(1, 23) = 26.86, p < .001$, reflected larger amplitudes for inverted than upright bodies (*Mdiff* = 1.00 μV). No other main effects or interactions were found, $ps > .05$.

Finally, a four-way interaction involving Site, Load, Inversion, and Distractor Type, $F(2, 46) = 4.59$, $p < .05$, simply reflected the fact that the above described effects were more pronounced at sites P9/P10 and PO9/PO10 than at P7/P8.

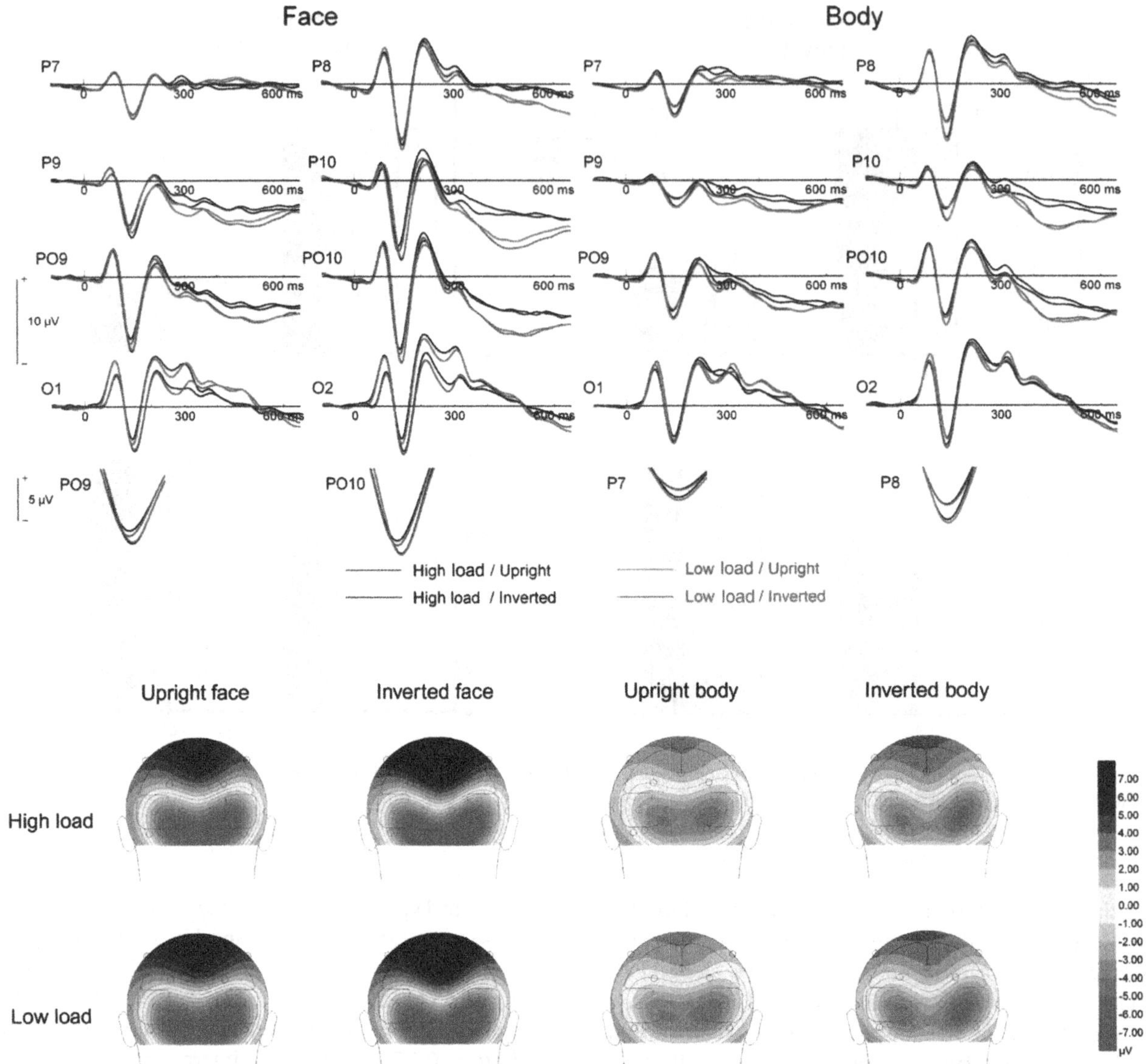

Figure 3. Grand average ERPs (top panel) in Experiment 1 for all conditions, shown separately for face and body distractors at posterior temporal and occipital sites. Enlarged visualization of prominent N170 peaks is given in a middle row (note that due to topographical differences, different electrode pairs were selected for faces and bodies). Topographical voltage maps (bottom panel) for N170 mean amplitudes as used in statistical analyses. Open circles indicate electrode positions. All maps were obtained by spherical spline interpolation, and at 90° equidistant projection. Negativity is red.

Late negative component (LNC). The late posterior negativity was quantified by using mean amplitudes between 350 and 600 ms at the same occipito-temporal sites used for the N170. An analogous ANOVA revealed a main effect of Hemisphere, $F(1, 23) = 7.35, p < .05$, with larger negativity over the right ($M = -2.57$ µV) than left hemisphere ($M = -1.49$ µV). LNC was larger for faces ($M = -2.49$ µV) than bodies ($M = -1.57$ µV), $F(1, 23) = 54.29, p < .001$, and larger under low than high load (*Mdiff* = 1.28 µV), $F(1, 23) = 60.75, p < .001$. There were additional interactions of Hemisphere by Load, $F(1, 23) = 19.82, p < .001$; Distractor Type by Load, $F(1, 23) = 5.01, p < .05$; and Distractor Type by Inversion, $F(1, 23) = 7.03, p < .01$.

The Load effect was reliable over both hemispheres—$t(23) = 7.47, p < .001$ and $t(23) = 4.70, p < .01$ for right and left hemisphere, respectively—but was larger over the right (*Mdiff* = 1.81 µV) than the left hemisphere (*Mdiff* = 0.72 µV). Load effects were significant for both distractor types—$t(23) = 7.96, p < .001$, and $t(23) = 6.38, p < .001$, for faces and bodies, respectively—but were slightly larger for faces (*Mdiff* = 1.43 µV) than bodies (*Mdiff* = 1.12 µV). Inversion effects were significant for

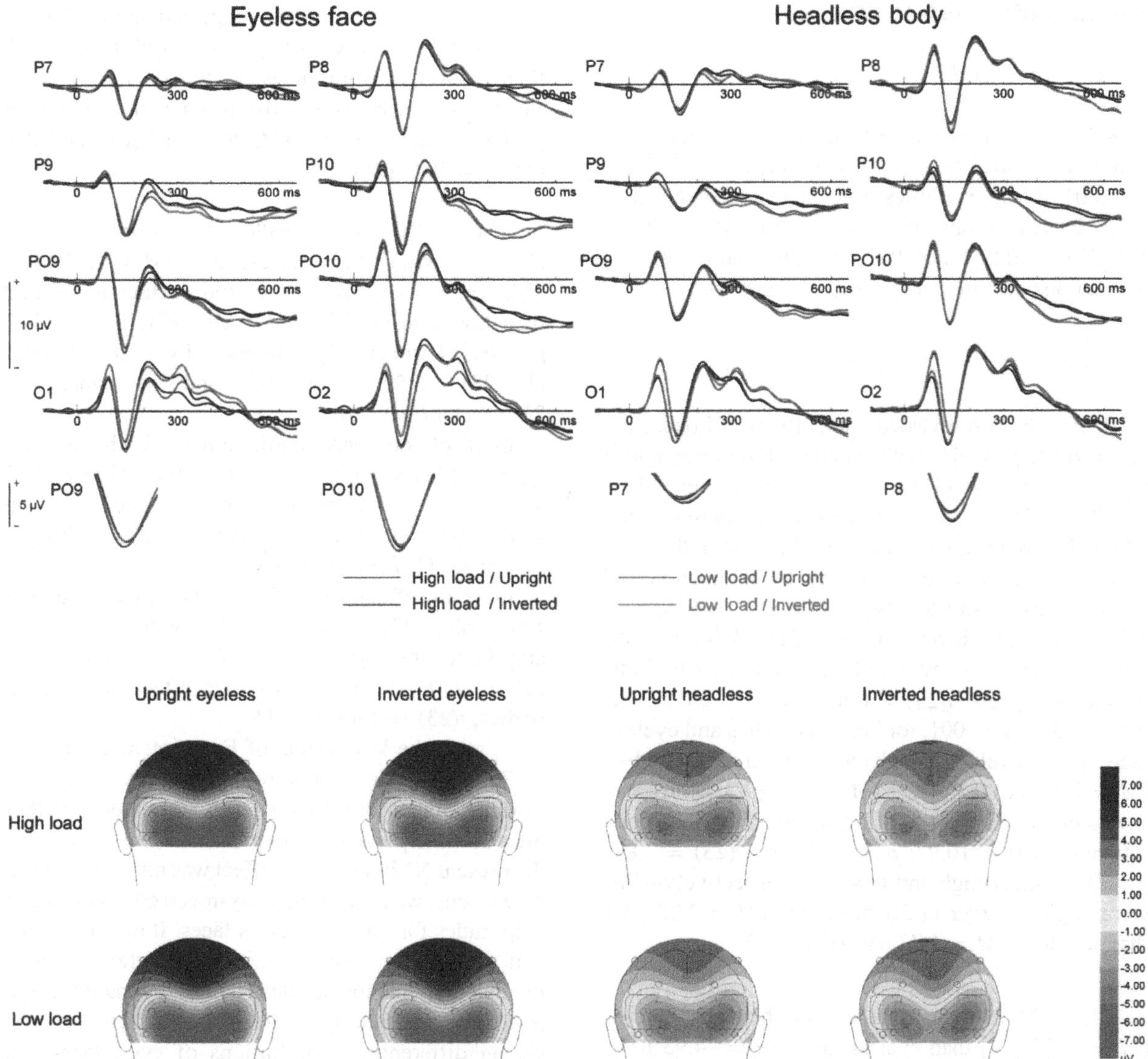

Figure 4. Grand average ERPs (top panel) and topographical voltage maps (bottom panel) in Experiment 2, shown separately for eyeless face and headless body distractors. All other parameters as in Figure 3.

bodies, $t(23) = 2.35, p < .05$, with larger amplitudes for inverted than upright bodies (*Mdiff* = 0.23 µV), but no such differences were seen for faces (*Mdiff* = 0.09 µV), $t(23) = 0.67, p > .20$.

Overall, load was successfully manipulated, with no evidence for any influence of task-irrelevant distractors on performance. By contrast, type and orientation of task-irrelevant distractors systematically modulated ERPs. Inversion generally increased P100 irrespective of load, and also elicited a load-dependent increase in N170, due to the fact that high load reduced N170 for upright faces only (Mohamed et al., 2009).

EXPERIMENT 2

Materials and methods

The same participants as in Experiment 1 were tested after a short interval of ~5 min, during which they completed the handedness questionnaire. Methods were identical to Experiment 1, and the same stimuli were used except that we removed the head region from body stimuli and the eye region (including eyebrows) from face stimuli. Data analyses were performed as in Experiment 1, using exactly the same dependent variables.

Results and discussion

Behavioral results

RTs were increased under high (M = 758 ms) compared to low load (M = 519 ms), $F(1, 23) = 117.18$, $p < .001$. Also, error rates were increased under high (M = 0.811) compared to low load (M = 0.971), $F(1, 23) = 102.77$, $p < .001$. No other main effects or interactions occurred, all $ps > .05$ (Figure 2).

ERPs

P100. ANOVA revealed main effects of Load, $F(1, 23) = 6.54$, $p < .01$, with smaller amplitudes under high than low load (*Mdiff* = 0.26 μV), and Inversion, $F(1, 23) = 107.48$, $p < .001$, with larger amplitudes for inverted than upright (*Mdiff* = 1.82 μV) stimuli.

These effects were moderated by interactions of Inversion by Distractor Type, $F(1, 23) = 9.76$, $p < .01$, and Load by Inversion, $F(1, 23) = 4.30$, $p < .05$. Follow-up analysis revealed Inversion effects for both distractor types—$t(23) = 8.87$, $p < .001$, and $t(23) = 5.11$, $p < .001$, for headless bodies and eyeless faces, respectively—though they were larger for bodies (M = 2.40 μV) than faces (M = 1.14 μV). Moreover, Inversion effects were significant under both load conditions—$t(23) = 10.97$, $p < .001$, and $t(23) = 7.89$, $p < .001$, under high and low load, respectively—but were slightly larger under high load (M = 2.00 μV) than low load (M = 1.73 μV) (Figure 4).

N170. N170 was smaller for headless bodies (M = –2.87 μV) than eyeless faces (M = –4.93 μV), $F(1, 23) = 90.18$, $p < .001$. Additionally, two-way interactions of Site by Distractor Type, $F(2, 46) = 33.84$, $p < .001$; Inversion by Distractor Type, $F(1, 23) = 16.72$, $p < .001$; and a four-way interaction of Hemisphere, Site, Inversion, and Distractor Type, $F(2, 46) = 3.91$, $p < .05$, were found. Regarding the two-way interactions, separate pair-wise t-tests for bodies and faces revealed an Inversion effect with larger N170 for inverted (M = –3.03 μV) than upright (M = –2.71 μV) bodies, $t(23) = 2.19$, $p < .05$. In contrast, N170 was larger to upright (M = –5.16 μV) than inverted (M = –4.70 μV) eyeless faces, $t(23) = -2.51$, $p < .05$ (Figure 4).

To resolve the four-way interaction, separate ANOVAs were conducted for headless bodies and eyeless faces. For bodies, Inversion interacted with Site, $F(2, 46) = 4.56$, $p < .05$, reflecting the largest Inversion effects at the more anterior electrode pair P7/P8 (*Mdiff* = 0.81 μV). For eyeless faces, a three-way interaction of Hemisphere, Site, and Inversion, $F(2, 46) = 3.52$, $p < .05$, suggested that—although the general pattern with larger amplitudes for upright than inverted stimuli was found throughout—Inversion effects were most prominent at the more posterior electrodes P10 (*Mdiff* = 0.66 μV), and PO9 (*Mdiff* = 0.77 μV).

LNC. LNC was smaller for headless bodies (M = –1.69 μV) than eyeless faces (M = –2.28 μV), $F(1, 23) = 24.56$, $p < .001$, and smaller under high than low load (*Mdiff* = 0.89 μV), $F(1, 23) = 40.83$, $p < .001$. Distractor Type interacted with both Load, $F(1, 23) = 6.59$, $p < .05$, and Inversion, $F(1, 23) = 11.17$, $p < .01$.

Load effects were significant for both distractor types—$t(23) = 5.04$, $p < .001$, and $t(23) = 7.05$, $p < .001$, for bodies and faces, respectively—but were slightly smaller for bodies (*Mdiff* = 0.76 μV) than faces (*Mdiff* = 0.99 μV).

Inversion effects on LNC amplitudes were found for faces only, $t(23) = -2.28$, $p < .05$, with slightly larger amplitudes for upright (M = –2.37 μV) than inverted (M = –2.18 μV) faces. No such effect occurred for bodies, $t(23) = 1.44$, $p > .15$.

In sum, the key results of Experiment 2 were that N170 inversion effects were dramatically reduced for headless bodies and eyeless faces, compared with their intact counterparts in Experiment 1. For eyeless faces, the present N170 inversion effect was not simply eliminated, but was even polarity-reversed, with larger amplitudes for upright eyeless faces. It may be important to note the more anterior distribution of N170 inversion effects for headless bodies compared to eyeless faces, particularly when considering current models on differential contributions of eye-, face-, and object- (or body-)selective neurons to N170 inversion effects (e.g., Itier et al., 2007). For both P100 and LNC components, the general pattern of results was very similar to that seen in Experiment 1, with prominent P100 modulations caused by inversion, and prominent LNC modulations caused by load.

GENERAL DISCUSSION

We measured prominent posterior P100, N170, and LNC components to investigate effects of load and inversion in perceiving body or face distractors. Relative to faces, the body N170 was smaller and had a more anterior temporal topography. Similar findings were reported by Thierry et al. (2006), who measured ERPs to task-relevant bodies and faces in a "one-back" repetition detection task. However, we did not replicate the ~20-ms delay for a body "N190" vs. a face N170 (see online supplementary material) reported by Thierry et al. (2006). Although this discrepancy could in part reflect task differences, we believe it is more likely that longer latencies in Thierry et al. (2006) were related to their incomplete (cropped) body images, particularly because a more recent study (Minnebusch et al., 2009) reported N170 at 168 ms to whole-body images. The topographies of the present N170 to bodies vs. faces indicate that both responses are generated by non-identical cortical sources, presumably related to the location of body- vs. face-sensitive cortical areas (Peelen & Downing, 2007). Here, N170 to bodies was distributed more anteriorly, while face N170 was prominent at posterior sites in Experiment 1. This topography may also relate to category-specific areas identified by transcranial magnetic stimulation (TMS) as crucial for face (right occipital face area) or body discriminations (right extrastriate body area; Pitcher, Charles, Devlin, Walsh, & Duchaine, 2009).

In Experiment 1, we observed larger N170 for inverted vs. upright images of both bodies and faces. This clearly demonstrates that such inversion effects can occur for task-irrelevant bodies and faces, in a similar way as reported previously for task-relevant stimuli (cf. Minnebusch et al., 2009; Stekelenburg & de Gelder, 2004). Moreover, N170 inversion effects were clear under high load, and even slightly enhanced for faces under high load. N170 inversion effects are often thought to reflect either disruption of holistic or configural processing (e.g., Minnebusch et al., 2009; Stekelenburg & de Gelder, 2004), increased difficulty of encoding, or a release of featural (e.g., eye) neurons from inhibition by holistic (e.g., face) neurons (Itier et al., 2007). Here we present the novel finding that N170 inversion effects not only occur for task-irrelevant stimuli, but also are relatively robust to attentional load. This shows that whichever interpretation of N170 inversion effects is correct, the process(es) that are reflected in this ERP and are altered by stimulus inversion are relatively independent of attention.

Experiment 2 exhibited an intriguingly different pattern of N170 inversion effects, for both headless bodies and eyeless faces.[1] For headless bodies, the N170 inversion effect was qualitatively similar but smaller compared to intact bodies. For eyeless faces, an "inverted inversion effect" occurred with smaller responses to inverted stimuli. With reference to eyeless faces, while an absence of the usual N170 increase by inversion is reminiscent of earlier findings by Itier et al. (2007), no "inverted inversion effect" was seen in that paper. Within their model, our findings could be explained by assuming reduced activity of face-selective neurons for inverted vs. upright (eyeless) faces. Although this assumption appears plausible, it was not made by Itier and Batty (2009), possibly to account for their equivalent N170 amplitudes for upright and inverted eyeless faces (Itier et al., 2007). The reason for this discrepancy with the present findings is unclear, but one obvious difference between studies is that Itier et al. (2007) used vertically cropped faces (without upper forehead and lower chin regions), perhaps to equate to the overall shape of object categories. Such cropping may have reduced activation of face neurons, which do seem to be sensitive to face orientation, as suggested by the present inverted inversion effects for eyeless faces. Future research may address this possibility more specifically. Visual encoding of bodies and faces was sometimes thought to recruit similar processes (Minnebusch & Daum, 2009; Peelen & Downing, 2007; Stekelenburg & de Gelder, 2004). Accordingly, aspects of the model by Itier and Batty (2009) might be applicable to body perception. Thus, the increased N170 by inverted intact bodies could partially reflect additional recruitment of head-sensitive cells, which normally receive inhibition by an upright body context. Our finding of a reduced N170 inversion effect for headless vs. intact bodies could be preliminary evidence for this idea, which receives independent support from a recent report of an "inverted N170 inversion effect" to headless bodies (Minnebusch et al., 2009). We tentatively suggest that inversion modulates the interactive encoding of hierarchical cues (body, head, eyes). As a caveat, and despite the appeal of this idea, a more comprehensive model will require a substantial degree of systematic future research (see also Minnebusch & Daum, 2009).

Whereas early ERPs were much more sensitive to inversion than to attentional load, the opposite was found for the LNC. In both experiments, LNC was reduced under high load across distractors (although the load effect was slightly larger for faces than bodies in both experiments). Due to topography and response to experimental variation, we interpret the LNC as a member of the family of posterior processing negativities previously observed both from scalp recordings (Czigler & Csibra, 1990; Schweinberger, Sommer, & Stiller, 1994) and from intracranial recordings at the ventral temporal cortical surface (Engell & McCarthy, 2010). We suggest that the present LNC indicates processing available for, and involuntarily (Lavie, 2005) allocated to, task-irrelevant distractors. Finally, the P100 was increased by inversion at occipital sites, as has also been reported by Itier and Taylor (2004). In line with that study, the present P100 inversion effect

[1]For a statistical comparison of the results of Experiments 1 and 2, see the online supplementary material, available via the 'Supplementary' tab on the article's online page (http://dx.doi.org/10.1080/17588928.2011.597848).

was strictly limited to occipital electrodes, whereas a number of studies failed to show a comparable P100 inversion effect when measuring amplitude at more anterior sites (Jacques & Rossion, 2007; see also Minnebusch et al., 2010).

REFERENCES

Bentin, S., Allison, T., Puce, A., Perez, E., & McCarthy, G. (1996). Electrophysiological studies of face perception in humans. *Journal of Cognitive Neuroscience*, *8*, 551–565.

Biederman, I. (1987). Recognition-by-components – a theory of human image understanding. *Psychological Review*, *94*, 115–147.

Czigler, I., & Csibra, G. (1990). Event-related potentials in a visual-discrimination task – negative waves related to detection and attention. *Psychophysiology*, *27*, 669–676.

Downing, P. E., Bray, D., Rogers, J., & Childs, C. (2004). Bodies capture attention when nothing is expected. *Cognition*, *93*, B27–B38.

Engell, A. D., & McCarthy, G. (2010). Selective attention modulates face-specific induced gamma oscillations recorded from ventral occipitotemporal cortex. *Journal of Neuroscience*, *30*, 8780–8786.

Itier, R. J., Alain, C., Sedore, K., & McIntosh, A. R. (2007). Early face processing specificity: It's in the eyes! *Journal of Cognitive Neuroscience*, *19*, 1815–1826.

Itier, R. J., & Batty, M. (2009). Neural bases of eye and gaze processing: The core of social cognition. *Neuroscience and Biobehavioral Reviews*, *33*, 843–863.

Itier, R. J., & Taylor, M. J. (2004). Effects of repetition learning on upright, inverted and contrast-reversed face processing using ERPs. *NeuroImage*, *21*, 1518–1532.

Jacques, C., & Rossion, B. (2007). Early electrophysiological responses to multiple face orientations correlate with individual discrimination performance in humans. *NeuroImage*, *36*, 863–876.

Jenkins, R., Lavie, N., & Driver, J. (2005). Recognition memory for distractor faces depends on attentional load at exposure. *Psychonomic Bulletin & Review*, *12*, 314–320.

Langton, S. R. H., Law, A. S., Burton, A. M., & Schweinberger, S. R. (2008). Attention capture by faces. *Cognition*, *107*, 330–342.

Lavie, N. (2005). Distracted and confused?: Selective attention under load. *Trends in Cognitive Sciences*, *9*, 75–82.

Maurer, D., Le Grand, R., & Mondloch, C. J. (2002). The many faces of configural processing. *Trends in Cognitive Sciences*, *6*, 255–260.

Minear, M., & Park, D. C. (2004). A lifespan database of adult facial stimuli. *Behavior Research Methods, Instruments, & Computers*, *36*, 630–633.

Minnebusch, D. A., & Daum, I. (2009). Neuropsychological mechanisms of visual face and body perception. *Neuroscience and Biobehavioral Reviews*, *33*, 1133–1144.

Minnebusch, D. A., Keune, P. M., Suchan, B., & Daum, I. (2010). Gradual inversion affects the processing of human body shapes. *NeuroImage*, *49*, 2746–2755.

Minnebusch, D. A., Suchan, B., & Daum, I. (2009). Losing your head: Behavioral and electrophysiological effects of body inversion. *Journal of Cognitive Neuroscience*, *21*, 865–874.

Mohamed, T. N., Neumann, M. F., & Schweinberger, S. R. (2009). Perceptual load manipulation reveals sensitivity of the face-selective N170 to attention. *Neuroreport*, *20*, 782–787.

Neumann, M. F., Mohamed, T. N., & Schweinberger, S. R. (2011). Face and object encoding under perceptual load: ERP evidence. *NeuroImage*, *54*, 3021–3027.

Nummenmaa, L., & Calder, A. J. (2009). Neural mechanisms of social attention. *Trends in Cognitive Sciences*, *13*, 135–143.

Peelen, M. V., & Downing, P. E. (2007). The neural basis of visual body perception. *Nature Reviews Neuroscience*, *8*, 636–648.

Perrett, D. I., Hietanen, J. K., Oram, M. W., & Benson, P. J. (1992). Organization and functions of cells responsive to faces in the temporal cortex. *Philosophical Transactions of the Royal Society of London, Series B: Biological Sciences*, *335*, 23–30.

Pitcher, D., Charles, L., Devlin, J. T., Walsh, V., & Duchaine, B. (2009). Triple dissociation of faces, bodies, and objects in extrastriate cortex. *Current Biology*, *19*, 319–324.

Reed, C. L., Stone, V. E., Bozova, S., & Tanaka, J. (2003). The body-inversion effect. *Psychological Science*, *14*, 302–308.

Ro, T., Friggel, A., & Lavie, N. (2007). Attentional biases for faces and body parts. *Visual Cognition*, *15*, 322–348.

Rossion, B., Delvenne, J. F., Debatisse, D., Goffaux, V., Bruyer, R., Crommelinck, M., et al. (1999). Spatio-temporal localization of the face inversion effect: An event-related potentials study. *Biological Psychology*, *50*, 173–189.

Rossion, B., Gauthier, I., Tarr, M. J., Despland, P., Bruyer, R., Linotte, S., et al. (2000). The N170 occipito-temporal component is delayed and enhanced to inverted faces but not to inverted objects: An electrophysiological account of face-specific processes in the human brain. *Neuroreport*, *11*, 69–74.

Sadeh, B., & Yovel, G. (2010). Why is the N170 enhanced for inverted faces? An ERP competition experiment. *NeuroImage*, *53*, 782–789.

Schweinberger, S. R., Sommer, W., & Stiller, R. M. (1994). Event-related potentials and models of performance asymmetries in face and word recognition. *Neuropsychologia*, *32*, 175–191.

Stekelenburg, J. J., & de Gelder, B. (2004). The neural correlates of perceiving human bodies: An ERP study on the body-inversion effect. *Neuroreport*, *15*, 777–780.

Tanaka, J. W., & Farah, M. J. (1993). Parts and wholes in face recognition *Quarterly Journal of Experimental Psychology, A, Human Experimental Psychology (London)*, *46*, 225–245.

Thierry, G., Pegna, A. J., Dodds, C., Roberts, M., Basan, S., & Downing, P. (2006). An event-related potential component sensitive to images of the human body. *NeuroImage*, *32*, 871–879.

Troje, N. F., & Westhoff, C. (2006). The inversion effect in biological motion perception: Evidence for a "life detector"? *Current Biology*, *16*, 821–824.

Yin, R. K. (1969). Looking at upside-down faces. *Journal of Experimental Psychology*, *81*, 141–145.

Yovel, G., Pelc, T., & Lubetzky, I. (2010). It's all in your head: Why is the body inversion effect abolished for headless bodies? *Journal of Experimental Psychology. Human Perception and Performance*, *36*, 759–767.

The influence of body-ownership cues on tactile sensitivity

Regine Zopf[1,2], Justin A. Harris[2], and Mark A. Williams[1]

[1]Macquarie Centre for Cognitive Science, Institute of Human Cognition and Brain Science, Macquarie University, Sydney, Australia
[2]School of Psychology, University of Sydney, Sydney, Australia

To some extent the bodies of others and one's own body are represented differently in the human brain. This study investigates how these different body representations are used during tactile perception. Two types of cues—purely visual cues (pictures of hands) and multisensory cues (equivalent to the rubber hand illusion paradigm)—were used to control whether a seen hand was one's own hand or somebody else's hand. We found that viewing one's own hand improves nonspatial tactile discrimination of supra-threshold stimuli, but attenuates tactile detection performance. Furthermore, when multisensory information signals that the viewed hand is not one's own hand, tactile nonspatial performance seems to be generally sensitized as compared to not viewing a hand. Such body-ownership-specific modulations were present only when multisensory cues signaled body ownership.

Keywords: Body ownership; Touch; Multisensory; Detection; Discrimination; Rubber hand illusion.

When we look at our own body, we naturally experience a strong sense of body ownership, which we do not experience when looking at other bodies. Indeed, brain activity in several areas has been shown to correlate with manipulations that signal body ownership. Purely visual information that signals body ownership, based on appearance and body orientation, is thought to be processed in the extrastriate body area (Myers & Sowden, 2008; Saxe, Jamal, & Powell, 2006). Synchronization of visual information to tactile sensations provides multisensory evidence that the seen body belongs to oneself, and this has been related to activity in multisensory areas such as the parietal, premotor, and insula cortices (Ehrsson, Holmes, & Passingham, 2005; Ehrsson, Spence, & Passingham, 2004; Tsakiris, Hesse, Boy, Haggard, & Fink, 2007).

The evidence for distinct subjective and neuronal correlates associated with looking at one's own body versus other bodies, raises important questions about if and how these representations are indeed used differently in order to guide and optimize perception that concerns our own body and others' bodies. More specifically, in the present studies, we investigated the effect of ownership for the seen body on unimodal tactile perception.

Previous research has shown that simply viewing a body site that is touched, even when there is no visual information about the external stimulus itself, improves spatial tactile discrimination in comparison to viewing nothing at all or viewing an object (Kennett, Taylor-Clarke, & Haggard, 2001). Further research has shown that viewing of a body site is not always beneficial. In fact, tactile detection and discrimination of near-threshold stimuli in a nonspatial task can be attenuated when viewing the body part, whereas tactile discrimination of supra-threshold stimuli in nonspatial tasks is improved (Harris, Arabzadeh, Moore, & Clifford, 2007). A possible explanation for these

Correspondence should be addressed to: Regine Zopf, Macquarie Centre for Cognitive Science, Macquarie University, Sydney, NSW 2109, Australia. E-mail: regine.zopf@mq.edu.au

R.Z. was supported by a Macquarie University Research Excellence Scholarship; J.A.H. by a Discovery Project Grant (DP0986137), and M.A.W. by a Queen Elizabeth II Fellowship (DP 0984919) from the Australian Research Council.

http://dx.doi.org/10.1080/17588928.2011.578208

findings is that vision of the body induces an adaptive change in tactile perception via a gain-control mechanism. More specifically, viewing a body may shift tactile sensitivity toward stimuli with higher amplitude and thus might optimize relative sensitivity for the purpose of discriminating between supra-threshold stimuli, at the expense of absolute sensitivity for the purpose of simple detection.

In the present study, we aimed to investigate whether a shift in gain-control occurs when viewing any body or only when viewing one's own body. We investigated the effect of purely visual information (Experiment 1) as well as of multisensory information (Experiments 2a and 2b) regarding body ownership. In both experiments, we included conditions in which no hand was seen that allowed us to investigate adaptive changes of viewing a specific hand relative to no view of a hand.

METHOD

Fifty-nine psychology students participated for course credit ($n = 11$ for Experiment 1; $n = 24$ for each of Experiments 2a and 2b). All participants gave written consent, and all experimental procedures were approved by the Human Research Ethics Committee at the University of Sydney.

Participants placed the pad of their left index finger on a tactile stimulator, which was placed inside a custom-made wooden frame (see Figure 1 for setup). Participants could not see their hand or the stimulator. Instead they saw an image presented on a monitor and reflected in a mirror. The monitor presented a recorded image of the inside of the wooden frame, containing the vibration stimulator. Depending on experimental condition, this was either a static or a video image. Care was taken to ensure that all images appeared to be at the same position and depth as the hand and stimulator.

We measured tactile thresholds for vibration amplitudes using a Bayesian adaptive staircase procedure (Kontsevich & Tyler, 1999). This procedure provides an optimal estimate of threshold within 30 trials, allowing us to keep the number of trials constant across conditions. Sensory thresholds were defined as 80.3% correct performance estimated from the fitted Weibull function with a 4% lapse rate.

Participants performed a two-interval, two-alternative, forced-choice task on each staircase trial. In Experiments 1 and 2a, detection thresholds were measured. The start of each 1500-ms interval was signaled by a short auditory cue (100 ms). One of the intervals contained a 500-ms, 40-Hz sinusoidal vibration that was presented 500 ms after the onset of one of the tones. Participants were instructed to indicate which interval contained the vibration. This response was made after they heard a third auditory cue that was presented at the very end of the trial. All auditory cues were presented through headphones that participants wore throughout the experiment, which also minimized any other sounds that were produced by the tactile device or, in Experiment 2, during brushing. Responses were performed with the right hand by pressing one of two mouse buttons. On each trial, the

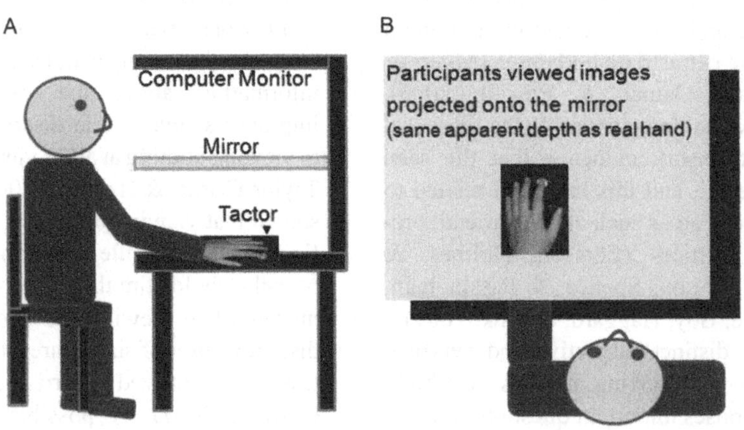

Figure 1. Experimental Setup. (A) View from side. (B) View from top. Participants placed the pad of their left index finger on a tactile stimulator ("tactor"), which was placed inside a custom-made wooden frame. The tactile stimulator consisted of a 3-mm-diameter steel rod driven by a vibration excitor and power amplifier (type 4810 "minishaker" and type 2718 amplifier; Brüel & Kjær, Nærum, Denmark). A brush was mounted above the participant's finger, and could be moved back and forth by a separate vibration excitor and power amplifier (type 4824 shaker and type 2732 amplifier; Brüel & Kjær, Nærum, Denmark). Both amplifiers were controlled by a computer running Matlab (Mathworks, Inc., Natick, MA, USA) with Psychtoolbox (Brainard, 1997; Pelli, 1997) via a National Instruments (Austin, TX, USA) interface card. A computer monitor was placed face down on top of the frame, and a mirror was mounted face up at a height exactly halfway between the tactile stimulator and the monitor.

staircase algorithm selects an amplitude of the to-be-detected stimulus so as to maximize the information gained after the response is given. For all experiments, the obtained sensory thresholds were normalized before statistical analysis; the threshold for each condition was divided by the participant's average threshold across all conditions.

The procedure in Experiment 2b was very similar to that used in Experiment 2a. The only difference was that discrimination thresholds were measured. A suprathreshold vibration was presented in both intervals on each trial, and participants were asked to indicate which of the two vibrations had the higher amplitude. The standard vibration was about 10 times higher than the average detection threshold obtained in pilot experiments.

Before each experiment, participants performed between two and five practice staircases without any visual stimuli. The threshold obtained on the last practice staircase was used to titrate for each participant the range of vibration amplitudes used in the staircases run in the main experiment.

In Experiment 1, participants viewed an image of their own hand, somebody else's hand (same gender), or the test setup without a hand. Participants viewed the hand of the previous participant with the same gender in the "other hand" condition. The first participants viewed the hand of the experimenter (R.Z.) in case of the first female participant and another male person in the lab in case of the first male participant. Note that participants always viewed the back of a hand, but were stimulated on the pad of their index finger (i.e., they did not see the exact skin surface that was stimulated). The image was displayed for 5 s before every trial (but not during the test trial), followed by a 500-ms gap before the cue that signaled the first interval of the test trial. Participants were instructed to look at the location of the fingernail of the index finger or the fixation cross which was positioned in place of the fingernail in conditions without a seen hand. Two staircases were obtained per condition, and the obtained threshold values were averaged. The order of the conditions was counterbalanced and presented in an ABCCBA fashion. Note that the counterbalancing was incomplete, as one of the six possible orders of conditions was only performed by one participant; all other orders were performed twice. After participants completed the first three conditions, we asked them to report in which condition they viewed their own hand. All participants reported this accurately.

In Experiments 2a and 2b, the influence of the synchrony of prior touch on tactile sensitivity was investigated. In order to modulate the sense of ownership, participants viewed a video of a brush stroking a hand while their own hidden hand was stroked by a brush for 2 min before the staircase was started. A 2 × 2 experimental design was implemented with the factors VIEW (view hand and view no hand) and STROKE (synchronous and asynchronous). The brush movement was automated and applied to the top side of the index finger. Each brush stroke lasted 1 s (0.5 s in each direction) and followed a sinusoidal movement of acceleration and deceleration, and a path length of about 3 cm. A 1-s stationary interval separated each stroke. In the synchronous condition, the movement of the brush in the video was synchronized to the movement of the brush on the participant's hand. In the asynchronous condition, a delay of 1 s was introduced between the seen and felt brush strokes, making them completely out of phase. We counterbalanced across participants whether the seen brushing was delayed or advanced by 1 s relative to the felt brushing. The staircase procedure was the same as in Experiment 1, except that participants received another five brush strokes between every trial. Again this was followed by a 500-ms gap before the tone that signaled the first interval of the threshold trial. Because this experiment included an additional condition, and the fact that each session included an extended period of brushing at the beginning of each staircase and further brushing between each test trial, time constraints meant that we could only reasonably run one staircase per condition (this took up to 1.5 h testing time for each participant). To compensate for the reduced amount of testing per condition, more participants were tested in Experiments 2a and 2b. The order of the four conditions was counterbalanced across participants.

After the staircase was run in a condition of Experiments 2a or 2b, participants were asked to rate seven statements to assess their subjective experience of ownership. The wording of the statements was based on previous questionnaires designed to assess the rubber hand illusion (Botvinick & Cohen, 1998; Longo, Schuur, Kammers, Tsakiris, & Haggard, 2008). We chose three specific statements that participants typically endorse more strongly when experiencing the rubber hand illusion, as well as four control questions. The rubber-hand-illusion rating statements were as follows: (1) "It seemed as if I were feeling the touch of the brush in the location where I saw the hand on the screen touched"; (2) "It seemed as though the touch I felt was caused by the brush touching the hand on the screen"; (3) "I felt as if the hand on the screen was my hand." In the *view no hand* conditions, the rating statements were adapted to account for the fact that there was no hand seen. For example, statement 3 was changed to "I felt as if my hand disappeared." The control statements in all conditions were as follows: (C1) "It seemed as if I might have more than one left hand or arm"; (C2) "I felt as if my (real) hand were turning 'rubbery'"; (C3) "I had the sensation that my hand was numb"; (C4) "It seemed like my hand was less vivid than normal." Participants rated each statement on a scale from 1 to 10, where 1 corresponded to "strongly disagree" and 10 to "strongly agree."

EXPERIMENT 1: INFLUENCE OF VISUAL CUES

Results and discussion

We removed from the analysis the data of two participants for which the obtained threshold values in at least one condition deviated more than 2 *SD* from the group mean. Across the remaining participants (mean age: 19.11 years, *SD*: 4.14; 66.6% female, 100% right-handed), detection thresholds ranged between 0.50 and 3.94 μm.

Figure 2 shows normalized thresholds for Experiment 1. Increased detection thresholds were observed for both conditions that involved viewing of a hand (either the participant's own hand or the other hand) compared to the condition with no hand. Two-tailed, paired *t*-tests confirmed the differences both for viewing one's own hand vs. no hand, $t(9) = 2.45$, $p = .04$, and viewing the other hand vs. no hand, $t(9) = 3.83$, $p = .005$. However, there was no significant difference in detection thresholds for viewing one's own hand and viewing the other person's hand, $t(9) = 0.44$, $p = .67$.

This finding is in line with a previous study that manipulated body ownership, using mainly visual cues (Haggard, 2006). Participants viewed either their own hand, a box placed above their own hand, or somebody else's hand which was placed on top of the box that covered their own hand while performing a spatial, grating-orientation discrimination task. The author found a significant difference in performance between the view of a hand (either the participant's own hand or the other hand) and the view of an object, but not between the view of one's own hand and the view of the other person's hand, consistent with our results here. Thus, it seems that both spatial and nonspatial tactile performance is unaffected by purely visual body-ownership cues for the seen hand.

Experiment 1 also replicated the previous finding that viewing a hand attenuates detection performance (Harris et al., 2007). However, the present setup is different from previous studies, in that participants could not actually see the body surface that received the tactile stimuli, but instead saw the other side of the stimulated finger. This indicates that the stimulated body site does not have to be viewed directly, and it is enough simply to view the stimulated body part. This is consistent with another recent study that found attenuation of tactile signal detection performance when viewing the hand from the top for tactile stimuli given to the finger pad (Mirams, Poliakoff, Brown, & Lloyd, 2010).

EXPERIMENTS 2A AND 2B: INFLUENCE OF SYNCHRONY OF PRIOR TOUCH

Results and discussion

We removed the data from two participants in Experiment 2a and one in Experiment 2b because their threshold values in at least one condition deviated more than 2 *SD* from the group mean. For Experiment 2a, across the remaining 22 participants (mean age: 20.7 years, *SD*: 4.41; 77.3% female, 100% right-handed), detection thresholds ranged between 0.60 and 3.97 μm. For Experiment 2b, across the remaining

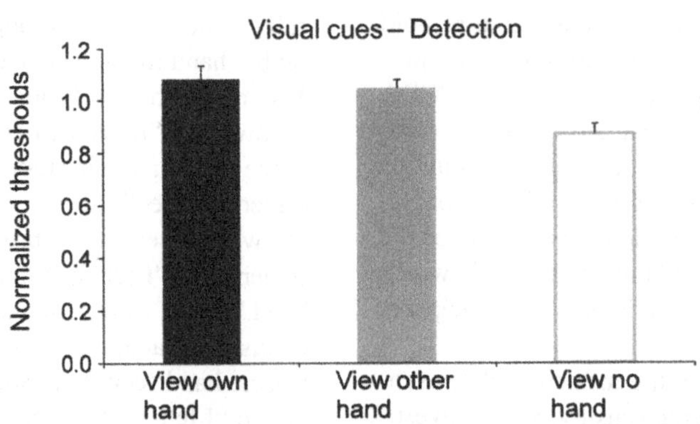

Figure 2. Influence of visual cues. Mean normalized detection thresholds for Experiment 1. Bars indicate 1 *SEM*.

23 participants (mean age: 22.3 years, *SD*: 4.35; 82.6% female, 95.7% right-handed), discrimination thresholds ranged between 1.74 and 7.38 μm.

We averaged responses for the three rating statements, which participants typically endorse when experiencing the rubber hand illusion (rating statements 1–3). We also combined the control rating statements (C1–C4). The averaged illusion ratings and control ratings are depicted in Figure 3. We tested whether the illusion rating mean was significantly different from the control rating mean for each of the four conditions, using two-tailed, paired *t*-tests. For both experimental groups, a significant difference was found for the condition in which a hand was seen combined with synchronous stroke—Experiment 2a, $t(21) = 3.17$, $p = .005$; Experiment 2b, $t(22) = 2.59$, $p = .017$—but not for any of the other conditions (all $p > .05$). This confirms that the sense of body ownership was induced when a hand was viewed and a synchronous brushing was applied.

The normalized thresholds for Experiment 2 are depicted in Figure 4. Viewing a hand with synchronized stroking increased detection thresholds (Experiment 2a) and lowered discrimination thresholds (Experiment 2b). In contrast, viewing a hand combined with asynchronous stroking generally lowered thresholds both for detection and discrimination. This pattern of results was statistically confirmed by 2 × 2 repeated-measurements ANOVAs with the factors VIEW (view hand vs. view no hand) and STROKE (synchronous vs. asynchronous). For Experiment 2a, neither of the main effects was significant—VIEW $F(1, 21) = 0.36$, $p = .55$; STROKE $F(1, 21) = 2.22$, $p = .15$—but we found a significant interaction between VIEW and STROKE: $F(1, 21) = 5.58$, $p = .028$. For Experiment 2b, we found a significant main effect of VIEW: $F(1, 22) = 5.64$, $p = .027$. Neither the main effect of STROKE, $F(1, 22) = .29$, $p = .60$, nor the interaction, $F(1, 22) = .56$, $p = .46$, was significant. When comparing the two body-ownership conditions for the seen hand directly, we found a significant difference between synchronous and asynchronous stroking for the view-hand condition in the detection task, $t(21) = 3.14$, $p = .005$, but not for the discrimination task, $t(22) = .139$, $p = .89$. Synchronous and asynchronous stroking were not significantly different in any of the view-no-hand comparisons (all $p > .05$).

The results presented here suggest a different pattern for viewing a hand combined with synchronous stroking as compared to asynchronous stroking. Viewing a hand with synchronous touch seems to attenuate tactile sensitivity in the detection task and improve discrimination performance, whereas viewing a hand with asynchronous touch seems to generally decrease threshold values as compared to not viewing a hand. To compare the performance patterns for detection and discrimination directly, we performed two additional ANOVAs for synchronous touch and asynchronous touch separately. We included the within-subject factor VIEW (view hand vs. view no hand) and the between-subject factor TASK (detection vs. discrimination). For synchronous touch, we found a significant interaction between VIEW and TASK, $F(1, 43) = 5.88$, $p = .02$, but no significant main effect for VIEW, $F(1, 43) = 0.153$, $p = .70$, whereas for asynchronous touch we found a significant main effect for VIEW, $F(1, 43) = 5.30$, $p = .026$, but no significant interaction between VIEW and TASK, $F(1, 43) = 0.814$, $p = .37$. These analyses thus support the notion of two

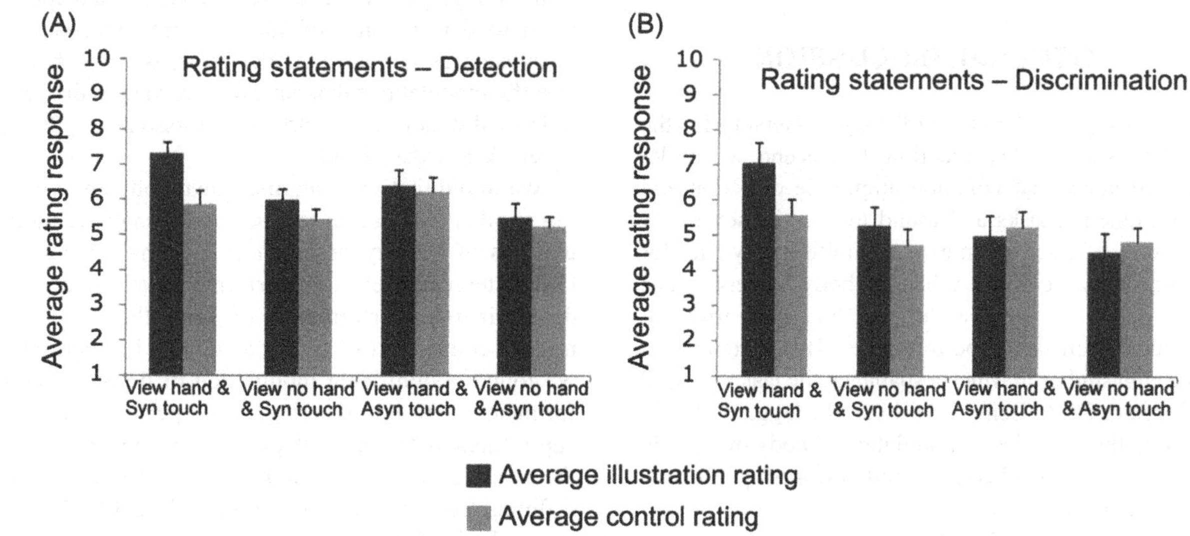

Figure 3. Rating statements results. Means for rubber hand illusion as well as control rating statements are given (A) for Experiment 2a – Detection, and (B) Experiment 2b – Discrimination. Bars indicate 1 *SEM*.

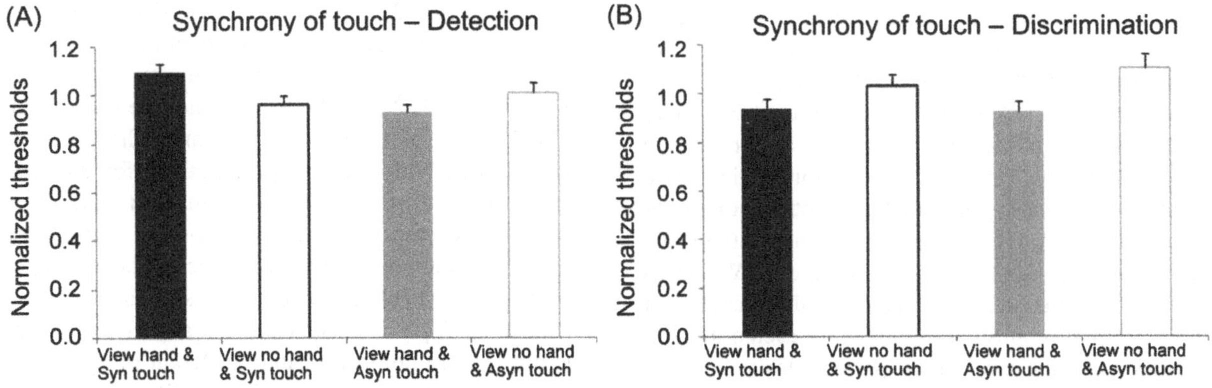

Figure 4. Influence of synchrony of touch. (A) Mean normalized detection thresholds for Experiment 2a. (B) Mean normalized discrimination thresholds for Experiment 2b. Bars indicate 1 *SEM*.

different patterns when multisensory information signals body ownership as compared to when it signals that the viewed body part belongs to somebody else.

To summarize, these results suggest that prior touch that signals body ownership for a seen hand leads to the same pattern of results that was previously found for just viewing the body without any previous touch: attenuation of tactile sensitivity for low-amplitude stimuli and improvement in discrimination for high-amplitude stimuli (Harris et al., 2007). However, when prior touch specifically signals that the seen body part does not belong to one's own body, because the viewed touch and the felt touch are asynchronous, tactile processing seems to be generally improved. These results imply that viewing a hand generally improves discrimination performance of supra-threshold stimuli, but there is a specific effect of hand ownership with respect to detection performance.

GENERAL DISCUSSION

We investigated the effect of body ownership for the seen body on tactile perception. To this end, we implemented nonspatial vibration amplitude detection and discrimination tasks and found that tactile sensitivity was only affected when explicit multisensory cues for ownership were given. When the body-ownership cue was purely visual, we did not find any effect of ownership on tactile performance. This pattern is in line with studies in which a spatial tactile task was used (Haggard, 2006; Longo, Cardozo, & Haggard, 2008). Also in these studies, a modulation of body ownership was only present when prior body ownership informative touch was given.

One possible confounding factor in our study is that the total viewing time for the body part differed between prior touch experiments (Experiments 2a and 2b) as compared to the purely visual experiment (Experiment 1). This was due to the longer period necessary to induce ownership for the seen hand. In Experiments 2a and 2b, participants saw the hand for 2 min at the beginning of each block and received five more brush strokes (amounting to 10 s) before every trial. In contrast, the hand was seen in Experiment 1 for only 5 s before every trial, and thus the viewing time was shorter than in Experiments 2a and 2b. We do not expect that a longer presentation of the seen hand would alter the pattern of results, as no significant effects of visual appearance on tactile perception were found in a previous study when the hand was seen for longer (Haggard, 2006).

It seems that for both spatial and nonspatial tactile perception, body ownership for a seen body part matters when prior multisensory information is presented. Thus, body-specific influences on tactile perception seem to depend on modulations from multisensory areas such as parietal bimodal areas, whereas body-specific modulations that have been related to the ventral visual stream are not effective (Ehrsson et al., 2004; Myers & Sowden, 2008).

We found that viewing one's own body and viewing another body seem to lead to very specific modulations of sensory processing. Viewing one's own body attenuates detection performance and improves discrimination performance of supra-threshold stimuli as compared to viewing an object. Furthermore, we found a significant modulation of detection performance, but not discrimination performance, of supra-threshold stimuli dependent on ownership of the seen hand. This finding seems to be somewhat different from findings by Longo et al. (2008). In that study, body ownership for a seen hand was also manipulated by the synchrony of touch, and spatial discrimination performance was tested. Because the task involved a fixed set of different grating ridge widths, performance for some participants was generally more accurate than for others. The authors divided the participants into two groups (high-performance and low-performance) and found an improvement of spatial discrimination after synchronous touch as compared to asynchronous touch only for participants who performed low overall. Holmes (2009) criticized this post-hoc method the authors used and pointed out that it is problematic because the result can be biased by regression toward the mean: "cells or subjects with low responses or poor performance in a 'unisensory' condition are likely to have higher responses or better performance in a multisensory condition, even if no 'real' multisensory integration occurs" (Holmes, 2009, p. 736). We avoided this statistical problem by using near-threshold stimuli in all participants. Furthermore, Longo et al. (2008) did not implement a no-view-hand condition; therefore, it is unclear whether spatial discrimination performance in both conditions is different from a condition where no hand is viewed. Below, we will discuss how the finding by Longo et al. (2008) could perhaps still be integrated with our results, discussing possible adaptive changes on the neuronal level in the somatosensory cortex.

Our findings with respect to viewing one's own body are in line with a shift in gain-control. According to this explanation, viewing one's own body optimizes tactile perceptual processing for stimuli with higher amplitude by reducing the gain for lower amplitude stimuli. This might be important for sensory processing during object manipulation and perception of moving stimuli that produce a sequence of high-amplitude signals. In contrast, when multisensory cues are asynchronous, tactile perception is generally improved, both for near-threshold and supra-threshold stimuli, as compared to viewing an object. Thus, it seems that, when multisensory information signals that the viewed body belongs to someone else, this generally sensitizes tactile perception. This might be related to previous findings which showed that viewing touch on another person without touch on one's own body activates somatosensory areas (Keysers et al., 2004), and viewing touch on another body improves tactile processing for near-threshold stimuli (Schaefer, Heinze, & Rotte, 2005; Serino, Pizzoferrato, & Ladavas, 2008). Such modulations might support processes involved in social perception and empathic understanding of others.

The present results are consistent with the suggestion that, when synchronous touch indicates that one is viewing one's own body, this leads to an adaptive shift in sensory gain toward higher-amplitude stimulation, as has been proposed for just viewing the body without explicit multisensory cues of ownership (Harris et al., 2007; Mirams et al., 2010). This would reduce absolute tactile detection sensitivity, but improve discrimination performance. Gain-control mechanisms in the visual modality have been related to changes in apparent receptive field size and changes in relative surround inhibition (Cavanaugh, Bair, & Movshon, 2002). Related changes in receptive field size could also account for differences between viewing one's own hand as compared to another hand in a spatial discrimination task (Longo, Cardozo, et al., 2008). Previously, changes in GABAergic intracortical inhibition have been proposed to account for changes in spatial discrimination performance and pain perception when viewing the body part (Cardini, Longo, & Haggard, 2011; Longo, Betti, Aglioti, & Haggard, 2009). Furthermore, Haggard et al. (2007) found that viewing the body changes the spatial gradient of masking effects on tactile spatial perception; the effect of distant maskers was reduced while the effect of close maskers was increased when viewing the body part as compared to an object. This finding supports the hypothesis that viewing the body might reduce tactile receptive field sizes possibly mediated by GABAergic intracortical and/or thalamocortical inhibition (Haggard, Christakou, & Serino, 2007). Such cortical and/or subcortical inhibitory mechanisms might thus change both neuronal gain and receptive field size and mediate the effects of viewing the body on nonspatial as well as spatial tactile and pain perception (Cardini et al., 2011; Haggard et al., 2007; Longo et al., 2009).

In sum, we found that viewing one's own body improves nonspatial tactile discrimination of supra-threshold stimuli and attenuates detection performance. This could be due to a shift in sensory gain. Furthermore, we found that multisensory information regarding touch that signals that the viewed hand is not one's own hand generally improves tactile nonspatial performance. Such body-ownership-specific modulations were only present when the synchrony of touch between the seen hand and the subject's own hand was manipulated. Thus, it seems that the brain makes use of body-ownership-signaling information regarding the synchrony of touch in order to optimize tactile perception.

REFERENCES

Botvinick, M., & Cohen, J. (1998). Rubber hands 'feel' touch that eyes see. *Nature, 391*(6669), 756.

Brainard, D. H. (1997). The Psychophysics Toolbox. *Spatial Vision, 10*(4), 433–436.

Cardini, F., Longo, M. R., & Haggard, P. (2011). Vision of the body modulates somatosensory intracortical inhibition. *Cerebral Cortex*. Advance online publication. doi: 10.1093/cercor/bhq267.

Cavanaugh, J. R., Bair, W., & Movshon, J. A. (2002). Nature and interaction of signals from the receptive field center and surround in macaque V1 neurons. *Journal of Neurophysiology, 88*(5), 2530–2546.

Ehrsson, H. H., Holmes, N. P., & Passingham, R. E. (2005). Touching a rubber hand: Feeling of body ownership is associated with activity in multisensory brain areas. *Journal of Neuroscience, 25*(45), 10564–10573.

Ehrsson, H. H., Spence, C., & Passingham, R. E. (2004). That's my hand! Activity in premotor cortex reflects feeling of ownership of a limb. *Science, 305*(5685), 875–877.

Haggard, P. (2006). Just seeing you makes me feel better: Interpersonal enhancement of touch. *Social Neuroscience, 1*(2), 104–110.

Haggard, P., Christakou, A., & Serino, A. (2007). Viewing the body modulates tactile receptive fields. *Experimental Brain Research, 180*(1), 187–193.

Harris, J. A., Arabzadeh, E., Moore, C. A., & Clifford, C. W. (2007). Noninformative vision causes adaptive changes in tactile sensitivity. *Journal of Neuroscience, 27*(27), 7136–7140.

Holmes, N. P. (2009). Inverse effectiveness, multisensory integration, and the bodily self: Some statistical considerations. *Consciousness and Cognition, 18*(3), 762–765.

Kennett, S., Taylor-Clarke, M., & Haggard, P. (2001). Noninformative vision improves the spatial resolution of touch in humans. *Current Biology, 11*(15), 1188–1191.

Keysers, C., Wicker, B., Gazzola, V., Anton, J. L., Fogassi, L., & Gallese, V. (2004). A touching sight: SII/PV activation during the observation and experience of touch. *Neuron, 42*(2), 335–346.

Kontsevich, L. L., & Tyler, C. W. (1999). Bayesian adaptive estimation of psychometric slope and threshold. *Vision Research, 39*(16), 2729–2737.

Longo, M. R., Betti, V., Aglioti, S. M., & Haggard, P. (2009). Visually induced analgesia: Seeing the body reduces pain. *Journal of Neuroscience, 29*(39), 12125–12130.

Longo, M. R., Cardozo, S., & Haggard, P. (2008). Visual enhancement of touch and the bodily self. *Consciousness and Cognition, 17*(4), 1181–1191.

Longo, M. R., Schuur, F., Kammers, M. P., Tsakiris, M., & Haggard, P. (2008). What is embodiment? A psychometric approach. *Cognition, 107*(3), 978–998.

Mirams, L., Poliakoff, E., Brown, R. J., & Lloyd, D. M. (2010). Vision of the body increases interference on the somatic signal detection task. *Experimental Brain Research, 202*(4), 787–794.

Myers, A., & Sowden, P. T. (2008). Your hand or mine? The extrastriate body area. *NeuroImage, 42*(4), 1669–1677.

Pelli, D. G. (1997). The VideoToolbox software for visual psychophysics: Transforming numbers into movies. *Spatial Vision, 10*(4), 437–442.

Saxe, R., Jamal, N., & Powell, L. (2006). My body or yours? The effect of visual perspective on cortical body representations. *Cerebral Cortex, 16*(2), 178–182.

Schaefer, M., Heinze, H. J., & Rotte, M. (2005). Viewing touch improves tactile sensory threshold. *Neuroreport, 16* (4), 367–370.

Serino, A., Pizzoferrato, F., & Ladavas, E. (2008). Viewing a face (especially one's own face) being touched enhances tactile perception on the face. *Psychological Science, 19* (5), 434–438.

Tsakiris, M., Hesse, M. D., Boy, C., Haggard, P., & Fink, G. R. (2007). Neural signatures of body ownership: A sensory network for bodily self-consciousness. *Cerebral Cortex, 17*(10), 2235–2244.

COGNITIVE NEUROSCIENCE, 2011, 2 (3–4), 155–162

Imagined paralysis impairs embodied spatial transformations

Matthias Hartmann, Caroline J. Falconer, and Fred W. Mast

Department of Psychology, Muesmattstrasse 45, University of Bern, Bern, Switzerland

Recent studies showed that motor deficits and limb amputations selectively impair mental rotation of respective body parts. This is due to modifications in the body schema, which plays a pivotal role in bodily related mental spatial transformations. In the present study, we investigated whether imagined paralysis could affect mental transformations in healthy participants. Participants were required to make leg laterality judgments of imitable and non-imitable body postures that were presented at different orientations. Mental spatial transformation of imitable body posture relies on emulation processes, a mechanism through which the posture is covertly imitated by the observer. Imagined paralysis selectively impaired mental transformation of imitable body postures. These results reflect an inability to fully emulate stimulus postures, suggesting a modulation in the body schema. Our results show that the body schema incorporates top-down information about motoric constraints which can influence embodied cognition in healthy participants.

Keywords: Embodiment; Imagery; Body schema; Mental transformation; Paralysis.

Metaphors are often used to facilitate abstract thinking. A specific form of metaphor is embodiment, a process through which abstract ideas are identified with a physical entity, namely the body (MacLachlan, 2004). Embodiment is a key cognitive component involved in some mental spatial transformation abilities. For example, to determine the handedness of a person who is writing, a mental representation of one's own body can be mentally aligned to match the physical composition of that person. A paradigm that is often used to investigate embodied spatial transformations requires participants to make laterality judgments of line drawings of human bodies or body parts (Parsons, 1987a, 1987b). Amorim, Isableu, and Jarraya (2006) have shown that performance in mental transformation tasks improves with increasing body-likeness of the stimuli. The authors distinguish between two types of embodiment processes. Spatial embodiment is the process of projecting and aligning our own body axes (front–back, top–bottom, left–right) to match those of the visual stimulus. Motoric embodiment refers to the engagement of the motor system covertly emulating the stimulus posture during mental body transformations (e.g., Grush, 2004; Parsons, 1994). This process is supported by findings that show partly overlapping brain activation when observing, imagining, or physically executing a body movement (e.g., Buccino et al., 2001; Decety & Grezes, 2006; Decety & Ingvar, 1990; Urgesi et al., 2010; Vogt, 1996).

Both embodiment processes rely on a mental representation of the body—the body schema (Head & Holmes, 1911–1912). The body schema is a multisensory construct that continuously updates the status of the body in space (Gallagher, 2005). The body schema

Correspondence should be addressed to: Matthias Hartmann, Department of Psychology, Muesmattstrasse 45, CH-3012 Bern, Switzerland. E-mail: matthias.hartmann@psy.unibe.ch

The study was funded by the Swiss National Science Foundation (Pro*Doc grant PDFMP1_127238 and Sinergia grant CRSII1-125135/1). We would like to thank Damaris Bachmann, Cora Bobst, and Nora Preuss for assistance in data collection.

http://dx.doi.org/10.1080/17588928.2011.594498

has long been conceived as a mere "body map," but recent studies have added to a more plastic, action-oriented construct (e.g., Berlucchi & Aglioti, 2010; Longo, Azanon, & Haggard, 2010; Maravita, Spence, & Driver, 2003). Investigations of tool use in humans and monkeys have shown that tools extending our physical boundaries can become incorporated into the body schema (see Maravita & Iriki, 2004, for a review). In contrast to body schema extension, there is also evidence for disintegration of body parts. Investigations with patients suffering from motor deficits due to hand dystonia (Fiorio, Tinazzi, & Aglioti, 2006; Fiorio et al., 2007) or body part amputation (Nico, Daprati, Rigal, Parsons, & Sirigu, 2004) suggest that the representation of the affected body part diminishes in the body schema. Nico et al. (2004) found that patients with upper limb amputation have difficulty in judging rotated hand stimuli as left or right. Furthermore, studies with healthy participants have shown that the actual body posture can influence the mental transformation of bodies or body parts (de Lange, Helmich, & Toni, 2006; Funk, Brugger, & Wilkening, 2005; Ionta & Blanke, 2009; Ionta, Fourkas, Fiorio, & Aglioti, 2007; Shenton, Schwoebel, & Coslett, 2004). These studies all suggest that the body schema relies on online integration of bottom-up proprioceptive signals. In this study, we investigate whether the body schema can be influenced by top-down processes via mental imagery in healthy participants. Mental imagery has been efficiently used to improve motor performance (Lotze & Halsband, 2006), muscle relaxation, and pain (e.g., Morone & Greco, 2007; Weydert et al., 2006). Specifically, we want to study whether the mental transformation of bodies could be influenced by an imagined immobility of one's own legs. Healthy participants were instructed to imagine themselves as paralyzed from the waist down while conducting leg laterality judgments of whole-body stimuli expressing imitable or non-imitable postures. Previous studies suggested that imitable postures and non-imitable postures are processed differently (Candidi, Urgesi, Ionta, & Aglioti, 2008; Cross, Mackie, Wolford, & de C. Hamilton, 2010). Amorim et al. (2006) suggested that imitable postures can be easily embodied, as the motor system can fully emulate the depicted posture. This is not the case for non-imitable postures. If imagined paralysis influences the body schema, we would expect an impairment of embodiment, at the stage of emulation, and consequently diminished performance in the mental transformation of imitable postures. Emulation processes cannot be applied successfully for non-imitable postures and, therefore, we would not expect effects of imagined paralysis. In a control task, participants were asked to perform spatial judgments on rotated objects. No embodiment processes are involved in spatial transformation of objects. We therefore expect no effect of imagined paralysis for the control task.

METHOD

Participants

Forty-four undergraduate students participated in this study (nine male, three left-handed). The mean age was 24.5, ranging from 19 to 36. They received course credit for their participation. The study was approved by the ethics committee of the University of Bern and participants gave informed consent.

Visual stimuli

For the laterality judgment task, photographs of six different people (three male and three female) demonstrating a yoga-like posture were used as stimuli. All actors had one leg in a straight position while the other leg was bent (see Figure 1 for an example). In order to enable motoric embodiment processes, we chose

Figure 1. Example of an imitable (left) and non-imitable (right) body posture. All six actors took a similar posture.

postures that were easily imitable. For each of the six postures, we created a non-imitable version by changing a body part into a biomechanically impossible position, thus preventing full emulation (Amorim et al., 2006). Including a mirrored version of all imitable and non-imitable postures resulted in a total of 24 different body stimuli. In addition to the laterality judgment task, we used a control task that required spatial judgments of objects. For this task, letters (R, F, L) and numbers (2, 5, 7) were used as stimuli. The original and mirrored version of each object resulted in 12 different object stimuli. For all stimuli (body and object), the size was approximately 15 cm presented on a 17-inch computer screen. Each stimulus was rotated in the picture plane at five angles of rotation (0°, 45°, 90°, 135°, 180°).

Task and procedure

The experiment consisted of a body block (120 trials) and an object block (60 trials). The body block contained a short break after half of the trials were completed. Each participant performed both blocks under two different conditions: a paralysis imagery condition and a control condition. In the paralysis imagery condition, participants were instructed to imagine themselves as paralyzed from the waist down and not to move their legs. In order to emphasize the idea of being paralyzed, participants were seated on a wheelchair. Before starting a stimulus block, the experimenter carefully pushed them around an obstacle positioned in the middle of the room. Participants remained seated on the wheelchair throughout the experiment in order to help to maintain the imagery state. In the control condition, no imagery instructions were given, and participants were asked to walk around the obstacle before sitting in a normal chair and starting a stimulus block. The true purpose of the study was concealed by a cover story. At the beginning of the experiment, participants were told that the purpose of this study was to investigate a possible influence of imagined paralysis on time estimation. They were instructed to remember how much time had elapsed since the beginning of a specific time point and, when required, to report a time estimation. Time estimations were collected at the end of each stimulus block. Participants were instructed to sit still during the experiment. The order of the two conditions was counterbalanced across participants, and a short break separated the two sessions. In addition, the order of the stimulus blocks (body and object) was counterbalanced across participants. Within the blocks, stimuli were presented in random order. In order to assess the effectiveness of the paralysis instruction, a subgroup of the participants ($n = 8$) were asked to fill out a questionnaire after the experiment. They had to read statements pertaining to the ownership and awareness of their legs and arms, and rate on a 7-point Likert scale the extent to which they agreed with these statements. The questionnaire was completed after the imagery and the control condition (see Table 1).

Participants were seated in front of a computer screen at a distance of approximately 60 cm. Task

TABLE 1
Subjective reports about legs and arms during paralysis imagery and control condition

	M (SD)	
Question	*Paralysis imagery*	*Control*
Legs		
During the experiment, my legs felt useless.	4.63 (1.85)	1.38 (0.52)
My legs felt heavy during the experiment.	4.88 (1.77)	1.25 (0.46)
During the experiment, I had the impression that I could not use my legs as well as I'm used to.	5.13 (1.25)	1.13 (0.35)
During the experiment, my legs felt weak.	5.13 (1.64)	1.00 (0.00)
During the experiment, my legs felt strange, as if they did not belong to my body.	3.13 (1.96)	1.00 (0.00)
During the experiment, my legs felt paralyzed.	3.63 (0.92)	1.00 (0.00)
During the experiment, I had the impression that I could not have moved my legs if I had wanted to.	3.25 (1.28)	1.13 (0.35)
It was difficult to stand up when the experiment was finished.	3.88 (1.89)	1.00 (0.00)
Mean leg score	**4.20 (0.69)**	**1.11 (0.18)**
Arms		
During the experiment, I had the impression that I could no longer move my arms.	1.63 (0.74)	1.00 (0.00)
During the experiment, my arms felt paralyzed.	1.13 (0.35)	1.00 (0.00)
During the experiment, my arms felt strange, as if they did not belong to my body.	1.13 (0.35)	1.00 (0.00)
During the experiment, my arms felt useless.	1.25 (0.71)	1.25 (0.46)
My arms felt heavy during the experiment.	2.25 (1.28)	1.38 (0.74)
During the experiment, I had the impression that I could not use my arms as well as I'm used to.	1.63 (1.06)	1.13 (0.35)
During the experiment, my arms felt weak.	1.63 (1.41)	1.00 (0.00)
Mean arm score	**1.52 (0.80)**	**1.12 (0.15)**

Notes: Values indicate mean scores on a 7-point Likert scale where 1 means strongly disagree and 7 means strongly agree ($n = 8$).

instructions appeared on the screen prior to each stimulus block. In the body block, participants were asked to judge, as accurately and as quickly as possible, whether the bent leg was left or right in origin. In the object block, participants were asked to judge, as accurately and as quickly as possible, whether the object appeared in its original or mirror-image form. Participants responded by pressing a left or right key with the index finger of the left or right hand. A central fixation cross was presented on the screen for 300 ms, followed by the stimulus that remained until the participant responded. Each response was followed by a blank screen of 700 ms. For each task, 10 practice trials were provided. Practice trials were not identical to the stimuli in the stimulus blocks. We used Superlab 4 Software (Cedrus Corporation, San Pedro, CA) for stimulus presentation and response collection.

Data analysis

The mean response time (RT) and error rate (ER) were calculated for each participant for the different stimulus categories, angle of rotation, and condition. Two participants were excluded from further analysis because of high ERs (58% and 46%; more than 3 *SD* above the mean ER). For the analysis of RTs, data were log transformed to normalize their distribution (Ratcliff, 1993). Incorrect trials (11.02% of all trials) and RTs more than 2.5 *SD* below or above the mean of each stimulus category and rotation angle (1.31% of all trials) were excluded from the analysis. RTs and ERs were analyzed for the laterality judgment task and the control task by means of separate repeated-measures analyses of variance (ANOVA). For the laterality judgment task, the analysis included the variables condition (paralysis imagery vs. control), posture (imitable vs. non-imitable), and angle of rotation (0°, 45°, 90°, 135°, 180°). For the control task, the analysis included the variables condition (paralysis imagery vs. control) and angle of rotation (0°, 45°, 90°, 135°, 180°). Statistical analysis was performed using PASW Statistics 18 (SPSS Inc., Chicago, IL). All post hoc tests were Bonferroni-corrected.

RESULTS

Laterality judgment task

RTs and ERs for each angle of rotation and condition are illustrated in Figure 2. The analysis of log RTs revealed a significant main effect of posture, $F(1, 41) = 40.50$, $MSE = 0.005$, $p < .001$, and a significant interaction between posture and condition, $F(1, 41) = 6.20$, $MSE = 0.002$, $p = .017$. RTs for imitable postures were lower than for non-imitable postures ($M = 1447$, $SD = 732$ vs. $M = 1577$, $SD = 855$),

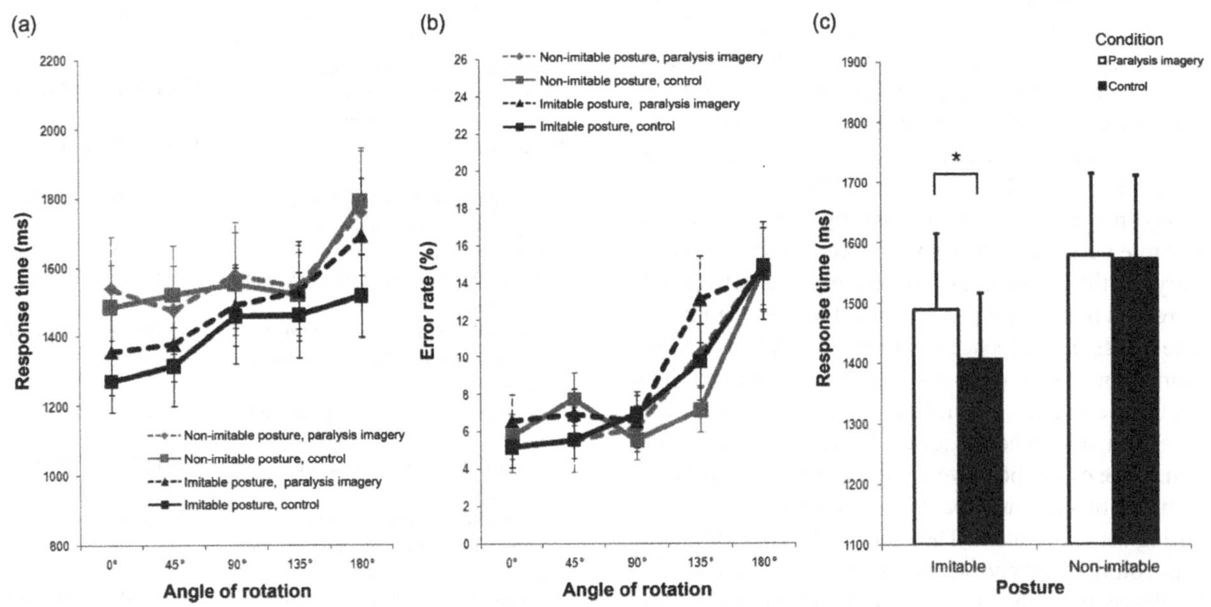

Figure 2. Mean response times (a) and error rates (b) for the laterality judgment task. Black lines represent imitable, and grey lines non-imitable body posture. Means are displayed at each angle of rotation during paralysis imagery condition (dotted line) and control condition (full line). Error bars indicate ±1 *SEM*. Interaction between posture and condition is illustrated in (c). Error bars indicate +1 *SEM*. Asterisk indicates a significant difference between the paralysis imagery condition and the control condition for the imitable postures ($p < .05$).

and paralysis imagery differentially affected these two types of stimuli. Post hoc tests revealed that imagined paralysis significantly increased RTs for the mental transformation of imitable postures, but no such effect was found for non-imitable postures (see Figure 2c). Moreover, the angle of rotation influenced RTs, $F(4, 164) = 21.83, MSE = 0.005, p < .001$. Correct spatial judgments required more time with increasing angle of rotation (see Figure 2a). Angle of rotation also interacted with posture, $F(4, 164) = 3.12$, $MSE = 0.002$, $p = .017$. Post hoc tests showed a significant difference between imitable and non-imitable postures for the rotation angles 0°, 45°, and 180°. All other effects or interactions were not significant ($p > .36$).[1] The analysis of ERs showed a main effect of angle of rotation, $F(4, 164) = 17.03$, $MSE = 2.00$, $p < .001$ (see Figure 2b). All other main effects or interactions were not significant ($p > .14$).

Object task

RTs and ERs for each angle of rotation and condition are illustrated in Figure 3. Analysis of log RTs revealed no difference between paralysis imagery and control condition, $F(1, 41) = 0.64$, $MSE = 0.013$. There was a significant effect only for angle of rotation, $F(4, 164) = 100.83, MSE = 0.006, p < .001$. Correct spatial judgements required more time with increasing angle of rotation (see Figure 3a). Analysis of ERs showed a main effect of degree of rotation, $F(4, 164) = 18.57$, $MSE = 2.578$, $p < .001$ (see Figure 3b). All other effects or interactions were not significant ($p > .85$).

Subjective reports and time estimations

The results of the subjective reports are shown in Table 1. Leg and arm scores were averaged for each participant and condition. The averaged scores were compared for both conditions by means of paired *t*-tests. During paralysis imagery, participants indicated significantly higher scores for the leg as compared to control condition, 4.20 vs. 1.12, $t(7) = 12.45$, $p < .001$. No significant difference between conditions was found for arm scores, 1.52 vs. 1.11, $t(7) = 1.38$, $p = .21$. This analysis shows that the paralysis imagery instruction had an effect on the subjective feeling of the participant's legs during the experiment.

The time estimations that were collected for the purpose of the cover story were averaged for each participant and condition. The duration of elapsed time was slightly overestimated in the paralysis imagery ($M = 12$ s, $SD = 79$ s) as well as in the control condition ($M = 15$ s, $SD = 108$ s). Paired *t*-tests showed no difference between the conditions, $t(41) = 0.31, p = .76$.

DISCUSSION

We examined whether imagined paralysis influences performance during mental spatial transformations of imitable and non-imitable body postures, and of objects. We found that leg laterality judgments of imitable body postures required more time during imagined paralysis as compared to the control condition. For postures that were not imitable, and for the object task, no effect of paralysis imagery was found. This shows that paralysis imagery exerted a selective influence on the processing of imitable body postures.

In order to correctly judge the laterality of body parts, participants imagine their own body in the orientation of the viewed posture (Parsons, 1994). If the posture was physically impossible to adopt, participants required in general more time for the laterality judgment. This shows that kinetic properties and biomechanical constraints are taken into consideration during mental spatial transformation (Overney & Blanke, 2009; Petit & Harris, 2005). The fact that there was no difference in ERs between imitable and non-imitable postures demonstrates that participants are able to solve the task with impossible postures (see also Craske, 1977; Moseley & Brugger, 2009; Romani, Cesari, Urgesi, Facchini, & Aglioti, 2005), but it demands more time. Interestingly, only imitable postures permitting full emulation were influenced by paralysis imagery. For non-imitable postures, where the emulation process is already impaired, performance during paralysis imagery

[1]Some studies report laterality effects as a function of handedness for mental spatial transformation of body parts (e.g., Nico et al., 2004). To investigate possible effects of laterality in the present data, the same analysis of RTs was performed with the additional between-variable laterality of the bent leg (left vs. right). This analysis shows that right-handed participants needed less time for correct judgments when the right leg was bent ($M = 1397$ ms, $SD = 579$ ms) as compared to the left leg ($M = 1427$ ms, $SD = 580$ ms), $F(1, 38) = 4.27, MSE = 0.009, p = .046$. This supports the observation by Nico et al. (2004) that participants use a strategy in which they initially mentally simulate movements of their dominant limb, which is the right leg in most right-handed people (Brown & Taylor, 1988). Laterality of the bent leg also interacted with angle of rotation, $F(4, 152) = 3.16, MSE = 0.005, p = .016$. Post hoc test showed that the difference between the left and right leg was significant only for 0°. All other interactions with laterality of the bent leg were not significant ($p > .12$). The small number of left-handed participants in our sample ($n = 3$) does not permit conclusions about the effect of handedness of the participants.

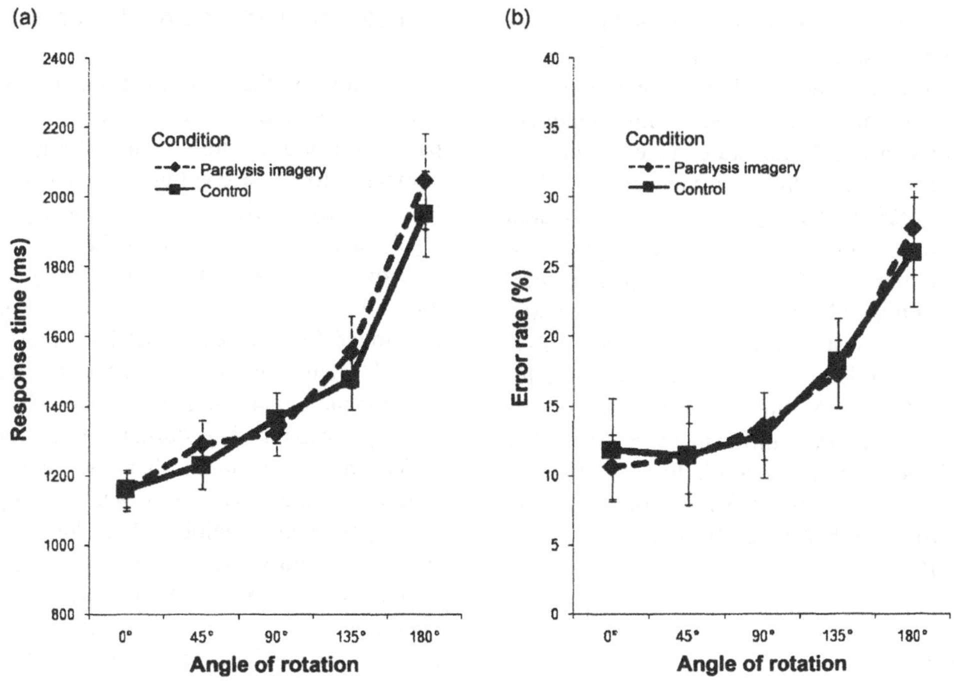

Figure 3. Mean response time (a) and error rate (b) for the control task. Means are displayed at each angle of rotation during paralysis imagery condition (dotted line) and control condition (full line). Error bars indicate ±1 *SEM.*

did not differ from the control condition. Our results suggest that paralysis imagery modulated the representation of the body, and this in turn impaired emulation processes during mental transformation of imitable postures. Imagined paralysis in healthy participants affected mental spatial transformations of the body in a way that is consistent with real motor deficits (Fiorio et al., 2006, 2007) or limb amputation (Nico et al., 2004). Therefore, these results extend our knowledge about the plasticity of the body schema by showing that some form of disintegration of body parts can be induced via top-down processes in healthy participants.

A frontoparietal network is involved in mental transformation of body stimuli (Creem et al., 2001; Vallar et al., 1999; Wraga, Shephard, Church, Inati, & Kosslyn, 2005; Zaehle et al., 2007). Patient studies have shown that lesions of this network, including the posterior parietal lobe (Overney & Blanke, 2009) and premotor cortex (Arzy, Overney, Landis, & Blanke, 2006), result in asomatognosia symptoms—the transient feeling that a body part has disappeared. Interestingly, these patients are impaired in mental transformations specific to body-related stimuli. The right insula is also associated with asomatognosia symptoms, and this brain area could be a core structure in maintaining a sense of limb ownership and awareness (Baier & Karnath, 2008; Karnath & Baier, 2010), as well as a coherent body representation (Karnath & Baier, 2010). There is also some evidence that the insula is involved in egocentric mental transformations (Wraga et al., 2005). Furthermore, some individuals deny the ownership of one or more healthy limbs and have a strong desire for amputation of that limb (apotemnophilia). This unusual disorder is characterized by disturbances in the integration of multisensory signals pertaining to the respective limb (Blanke, Morgenthaler, Brugger, & Overney, 2009; Hilti & Brugger, 2010) and even an exclusion of the limb from the body schema (Brang, McGeoch, & Ramachandran, 2008). Preliminary findings suggest that the right parietal lobe is the source of such alterations in corporeal awareness and ownership (Blanke et al., 2009; Brang et al., 2008; McGeoch et al., 2009). In light of the above, the premotor cortex, insula, and parietal areas are the prime candidates underlying the results we obtained in this study. On the one hand, lesions in these areas give rise to deficits in a coherent body representation; on the other hand, these areas are implicated in mental transformation of the body.

Modifying the body schema via mental imagery could be a promising approach for the treatment of clinical disorders involving dysfunctional body representations. For example, patients suffering from conversion paralysis are unable to perform voluntary motions despite the absence of motor deficits (de Vignemont, 2008). These patients also show selective motor imagery impairments for their affected limb (de Lange, Toni, & Roelofs, 2010). It has been suggested that conversion paralysis could result from an inability to form a "mental image" of movements (for a review, see Vuilleumier, 2005). For these patients, mental imagery or hypnotic instruction about their own corporeal awareness and motion potential could be used to improve motor performance. It would also be interesting to investigate whether it is possible to reduce symptoms in patients with asomatognosia, or people with a desire for amputation by modulating their body representation via mental imagery. An additional future study could also investigate whether imagined paralysis is accompanied by physiological changes in the "affected" limb. Moseley et al. (2008) demonstrated that altered body representation induced by conflicting visual-tactile input decreased skin temperature in the real hand.

To conclude, earlier studies have shown that modifying the body schema via bottom-up stimuli can influence the ability to mentally transform bodies or body parts (Ionta et al., 2007; Shenton et al., 2004). Our study is the first to highlight the influence of top-down information via mental imagery on the body schema and embodiment processes in healthy participants. The results demonstrate that the body schema is not a purely sensory-driven construct but is cognitively malleable via top-down processes (see also Moseley & Brugger, 2009). Top-down processes involved in one's own motoric constraints influence the body schema, and this in turn influences embodied cognition.

REFERENCES

Amorim, M. A., Isableu, B., & Jarraya, M. (2006). Embodied spatial transformations: "Body analogy" for the mental rotation of objects. *Journal of Experimental Psychology: General*, *135*(3), 327–347.

Arzy, S., Overney, L. S., Landis, T., & Blanke, O. (2006). Neural mechanisms of embodiment: Asomatognosia due to premotor cortex damage. *Archives of Neurology*, *63*(7), 1022–1025.

Baier, B., & Karnath, H. O. (2008). Tight link between our sense of limb ownership and self-awareness of actions. *Stroke*, *39*(2), 486–488.

Berlucchi, G., & Aglioti, S. M. (2010). The body in the brain revisited. *Experimental Brain Research*, *200*(1), 25–35.

Blanke, O., Morgenthaler, F. D., Brugger, P., & Overney, L. S. (2009). Preliminary evidence for a fronto-parietal dysfunction in able-bodied participants with a desire for limb amputation. *Journal of Neuropsychology*, *3*(Pt 2), 181–200.

Brang, D., McGeoch, P. D., & Ramachandran, V. S. (2008). Apotemnophilia: A neurological disorder. *Neuroreport*, *19*(13), 1305–1306.

Brown, E. R., & Taylor, P. (1988). Handedness, footedness, and eyedness. *Perceptual and Motor Skills*, *66*(1), 183–186.

Buccino, G., Binkofski, F., Fink, G. R., Fadiga, L., Fogassi, L., Gallese, V., et al. (2001). Action observation activates premotor and parietal areas in a somatotopic manner: An fMRI study. *European Journal of Neuroscience*, *13*(2), 400–404.

Candidi, M., Urgesi, C., Ionta, S., & Aglioti, S. M. (2008). Virtual lesion of ventral premotor cortex impairs visual perception of biomechanically possible but not impossible actions. *Social Neuroscience*, *3*(3–4), 388–400.

Craske, B. (1977). Perception of impossible limb positions induced by tendon vibration. *Science*, *196*(4285), 71–73.

Creem, S. H., Downs, T. H., Wraga, M., Harrington, G. S., Proffitt, D. R., & Downs, J. H., 3rd (2001). An fMRI study of imagined self-rotation. *Cognitive, Affective, & Behavioral Neuroscience*, *1*(3), 239–249.

Cross, E. S., Mackie, E. C., Wolford, G., & de C. Hamilton, A. F. (2010). Contorted and ordinary body postures in the human brain. *Experimental Brain Research*, *204*(3), 397–407.

Decety, J., & Grezes, J. (2006). The power of simulation: Imagining one's own and other's behavior. *Brain Research*, *1079*(1), 4–14.

Decety, J., & Ingvar, D. H. (1990). Brain structures participating in mental simulation of motor behavior: A neuropsychological interpretation. *Acta Psychologica*, *73*(1), 13–34.

de Lange, F. P., Helmich, R. C., & Toni, I. (2006). Posture influences motor imagery: An fMRI study. *NeuroImage*, *33*(2), 609–617.

de Lange, F. P., Toni, I., & Roelofs, K. (2010). Altered connectivity between prefrontal and sensorimotor cortex in conversion paralysis. *Neuropsychologia*, *48*(6), 1782–1788.

de Vignemont, F. (2008). Hysterical conversion: The reverse of anosognosia? In T. Bayne & J. Fernandez (Eds.), *Delusions and self-deception: Affective influences on belief-formation* (pp. 243–262). Hove, UK: Psychology Press.

Fiorio, M., Tinazzi, M., & Aglioti, S. M. (2006). Selective impairment of hand mental rotation in patients with focal hand dystonia. *Brain*, *129*(Pt 1), 47–54.

Fiorio, M., Tinazzi, M., Ionta, S., Fiaschi, A., Moretto, G., Edwards, M. J., et al. (2007). Mental rotation of body parts and non-corporeal objects in patients with idiopathic cervical dystonia. *Neuropsychologia*, *45*(10), 2346–2354.

Funk, M., Brugger, P., & Wilkening, F. (2005). Motor processes in children's imagery: The case of mental rotation of hands. *Developmental Science*, *8*(5), 402–408.

Gallagher, S. (2005). *How the body shapes the mind*. New York, NY: Oxford University Press.

Grush, R. (2004). The emulation theory of representation: Motor control, imagery, and perception. *Behavioral and Brain Sciences*, *27*(3), 377–396.

Head, H., & Holmes, G. (1911–1912). Sensory disturbances from cerebral lesions. *Brain*, *34*, 102–254.

Hilti, L. M., & Brugger, P. (2010). Incarnation and animation: Physical versus representational deficits of body integrity. *Experimental Brain Research*, *204*(3), 315–326.

Ionta, S., & Blanke, O. (2009). Differential influence of hands posture on mental rotation of hands and feet in left and right handers. *Experimental Brain Research*, *195*(2), 207–217.

Ionta, S., Fourkas, A. D., Fiorio, M., & Aglioti, S. M. (2007). The influence of hands posture on mental rotation of hands and feet. *Experimental Brain Research*, *183*(1), 1–7.

Karnath, H. O., & Baier, B. (2010). Right insula for our sense of limb ownership and self-awareness of actions. *Brain Structure & Function*, *214*(5–6), 411–417.

Longo, M. R., Azanon, E., & Haggard, P. (2010). More than skin deep: Body representation beyond primary somatosensory cortex. *Neuropsychologia*, *48*(3), 655–668.

Lotze, M., & Halsband, U. (2006). Motor imagery. *Journal of Physiology*, *99*(4–6), 386–395.

MacLachlan, M. (2004). *Embodiment: Clinical, critical and cultural perspectives on health and illness*. Maidenhead, UK: Open University Press.

Maravita, A., & Iriki, A. (2004). Tools for the body (schema). *Trends in Cognitive Sciences*, *8*(2), 79–86.

Maravita, A., Spence, C., & Driver, J. (2003). Multisensory integration and the body schema: Close to hand and within reach. *Current Biology*, *13*(13), R531–539.

McGeoch, P. D., Brang, D., Song, T., Lee, R. R., Huang, M., & Ramachandran, V. S. (2009). Apotemnophilia — the neurological basis of a 'psychological' disorder. *Nature Procedings*, hdl:10101/npre.12009.12954.10101.

Morone, N. E., & Greco, C. M. (2007). Mind—body interventions for chronic pain in older adults: A structured review. *Pain Medicine*, *8*(4), 359–375.

Moseley, G. L., & Brugger, P. (2009). Interdependence of movement and anatomy persists when amputees learn a physiologically impossible movement of their phantom limb. *Proceedings of the National Academy of Sciences of the United States of America*, *106*(44), 18798–18802.

Moseley, G. L., Olthof, N., Venema, A., Don, S., Wijers, M., Gallace, A., et al. (2008). Psychologically induced cooling of a specific body part caused by the illusory ownership of an artificial counterpart. *Proceedings of the National Academy of Sciences of the United States of America*, *105*(35), 13169–13173.

Nico, D., Daprati, E., Rigal, F., Parsons, L., & Sirigu, A. (2004). Left and right hand recognition in upper limb amputees. *Brain*, *127*, 120–132.

Overney, L. S., & Blanke, O. (2009). Impaired imagery for upper limbs. *Brain Topography*, *22*(1), 27–43.

Parsons, L. M. (1987a). Imagined spatial transformation of one's body. *Journal of Experimental Psychology: General*, *116*(2), 172–191.

Parsons, L. M. (1987b). Imagined spatial transformations of one's hands and feet. *Cognitive Psychology*, *19*(2), 178–241.

Parsons, L. M. (1994). Temporal and kinematic properties of motor behavior reflected in mentally simulated action. *Journal of Experimental Psychology: Human Perception and Performance*, *20*(4), 709–730.

Petit, L. S., & Harris, I. M. (2005). Anatomical limitations in mental transformations of body parts. *Visual Cognition*, *12*(5), 737–758.

Ratcliff, R. (1993). Methods for dealing with reaction time outliers. *Psychological Bulletin*, *114*(3), 510–532.

Romani, M., Cesari, P., Urgesi, C., Facchini, S., & Aglioti, S. M. (2005). Motor facilitation of the human corticospinal system during observation of bio-mechanically impossible movements. *NeuroImage*, *26*(3), 755–763.

Shenton, J. T., Schwoebel, J., & Coslett, H. B. (2004). Mental motor imagery and the body schema: Evidence for proprioceptive dominance. *Neuroscience Letters*, *370*(1), 19–24.

Urgesi, C., Maieron, M., Avenanti, A., Tidoni, E., Fabbro, F., & Aglioti, S. M. (2010). Simulating the future of actions in the human corticospinal system. *Cerebral Cortex*, *20* (11), 2511–2521.

Vallar, G., Lobel, E., Galati, G., Berthoz, A., Pizzamiglio, L., & Le Bihan, D. (1999). A fronto-parietal system for computing the egocentric spatial frame of reference in humans. *Experimental Brain Research*, *124*(3), 281–286.

Vogt, S. (1996). Imagery and perception-action mediation in imitative actions. *Cognitive Brain Research*, *3*(2), 79–86.

Vuilleumier, P. (2005). Hysterical conversion and brain function. *Progress in Brain Research*, *150*, 309–329.

Weydert, J. A., Shapiro, D. E., Acra, S. A., Monheim, C. J., Chambers, A. S., & Ball, T. M. (2006). Evaluation of guided imagery as treatment for recurrent abdominal pain in children: A randomized controlled trial. *BMC Pediatrics*, *6*, 29.

Wraga, M., Shephard, J. M., Church, J. A., Inati, S., & Kosslyn, S. M. (2005). Imagined rotations of self versus objects: An fMRI study. *Neuropsychologia*, *43*(9), 1351–1361.

Zaehle, T., Jordan, K., Wustenberg, T., Baudewig, J., Dechent, P., & Mast, F. W. (2007). The neural basis of the egocentric and allocentric spatial frame of reference. *Brain Research*, *1137*(1), 92–103.

COGNITIVE NEUROSCIENCE, 2011, 2 (3–4), 163–170

Differential effects of perceived hand location on the disruption of embodiment by apparent physical encroachment of the limb

Catherine Preston[1,2] and Roger Newport[2]

[1]Division of Psychology, Nottingham Trent University, Nottingham, UK
[2]School of Psychology, University of Nottingham, Nottingham, UK

While much is known about the processes underlying the embodiment of non-corporeal objects, less research has been conducted to explore the mechanisms that prevent embodiment. The current study investigates the effect of violation of the human body template on embodiment of fake limbs by appearing to encroach the fake limb with a solid object. Encroachment of the limb disrupted embodiment, but only when the limb was far from the body. When the encroached limb was close to the body, it was not disembodied despite such clear violations of knowledge related to the body. The representation of the limb in the body schema was similarly robust to near-hand violations, suggesting that competing representations of the limb that are close to the body are preferred to those that are far from the body.

Keywords: Body ownership; Body image; Body schema; Embodiment; Body violation.

Many experiments in recent years have investigated the embodiment of extra-corporeal objects into the body image (representation of the body for perception) and body schema (representation of the body for action). For example, rubber hands (Botvinick & Cohen, 1998), fake supernumerary limbs (Newport, Pearce, & Preston, 2010), tools (Kao & Goodale, 2009) and even tables (Armel & Ramachandran, 2003) have variously been shown to be felt, or treated, as being part of the body. However, there have been many fewer investigations of the disembodiment of actual body parts (for a review, see Giummarra, Gibson, Georgiou-Karistanis, & Bradshaw, 2008) and fewer still into the mechanisms that prevent embodiment of objects and body parts. Within neuropsychology, there are numerous, largely anecdotal examples of patients' failure to accept their own limb as part of themselves. Somatoparaphrenia is a disorder mainly associated with the right hemisphere that is characterized by the patient holding delusional beliefs about the contralesional side of the body, the most frequently reported delusion being disownership of a contralesional limb (*asomatognosia*) (for review, see Vallar & Ronchi, 2009).

We often find ourselves in situations in which our hand lies in close proximity to a non-bodily object, such as a cup or glass sitting next to us as we type on a computer keyboard, but despite its physical proximity to our own hand, it does not get mistaken for our hand and the question at issue here is, 'why not?' Some of the reasons are fairly clear: For example, there are substantial physical differences between the cup and the hand, and it fails to move in a manner predicted by our sensorimotor systems. When the physical and sensory input does match, however, non-bodily objects can be mistaken for our own body parts. For example, patient M.P., with a large parietal lesion, was unable to move his left hand following stroke, but if his wife's hand turned the page of a newspaper for him at the

Correspondence should be addressed to: Catherine Preston, Division of Psychology, School of Social Sciences, Nottingham Trent University, Chaucer Building, Goldsmith Street, Nottingham, NG1 5LT, UK. Email: catherine.preston@ntu.ac.uk

C.P. was funded by ESRC grant (RES-000-22-3455), and the authors would like to thank Brid Dineen for assistance with data collection.

http://dx.doi.org/10.1080/17588928.2011.582944

same time that he was thinking about turning the page, then he would mistake his wife's hand for his own (Preston & Newport, 2008). In healthy controls, with congruent sensory input (which can refer to spatial, visual, or temporal congruency), congruently moving fake hands (Newport et al., 2010) and even static rubber hands (Botvinick & Cohen, 1998) can feel as though they are our own.

While congruent sensory feedback helps to embody objects, the failure to embody might not simply be down to incongruent sensory feedback, for we do not mistake tools, which often do provide congruent sensory feedback, as part of our body. Peripersonal space may expand (Iriki, Tanaka, & Iwamura, 1996; Maravita, Spence, Kennett, & Driver, 2002), and it can be argued that such tools are incorporated into our body representations (Kao & Goodale, 2009), but they are not perceived as actual body parts. There may be a variety of factors that prevent the false embodiment of nearby objects such as if they violate our knowledge of our body; for example, embodiment of a synchronously touched fake limb (congruent sensory feedback) can be abolished if the fake limb is positioned in an implausible posture relative to the body (Farne, Pavani, Meneghello, & Ládavas, 2000; Pavani, Spence, & Driver, 2000; Tsakiris & Haggard, 2005). Even body parts that do not physically exist (e.g., phantom limbs and point light displays) tend to obey our real-world knowledge or our naive understanding of the laws of physics, and one of those laws is that bodies do not pass through solid objects. Even young infants know this: Human motion studies have shown that infants as young as 6 months know that human body parts and even moving human point light displays should not pass through solid objects (Moore, Goodwin, George, Axelsson, & Braddick, 2007; Saxe, Tzelnic, & Carey, 2006). In addition, many of those who experience phantom limbs with congenital aplasia report that their phantoms telescope or disappear to avoid passing through physical objects (e.g., Poeck, 1964), although there are examples of phantoms that will (e.g., Henderson & Smyth, 1948), and there are also cases of supernumerary phantom limbs that cannot penetrate solid objects (e.g., Khateb et al., 2009). Furthermore, when illusions are presented such that the arm is perceived as moving through a solid object as opposed to around it, brain activity switches from neural circuits associated with human motion to those more associated with the processing of objects and objects in motion (Stevens, Fonlupt, Shiffrar, & Decety, 2000).

The current experiment was designed to investigate whether violation of the physical space associated with the hand would be sufficient for the disownership and disembodiment of that hand. Participants were presented with two identical images of their own real hand (i.e., not rubber hands). Both moved synchronously with the their own movements (delay < 17 ms) and stroked a sponge-tipped stick for 20 s. At the end of the stroking period, the stick appeared to move through (encroach) the fake hand that was nearest to the body midline, the fake hand furthest from the body midline, or neither hand, and this was immediately followed by either ownership questionnaires or pointing to targets without vision of the real hand. Previous research demonstrates that when two fake hands both have synchronous multisensory feedback with the real hand, both are incorporated into the body image (Ehrsson, 2009; Newport et al., 2010). It was predicted that encroachment of either one of the hands would reduce ownership of that hand in terms of the body image, as measured by questionnaire, such that verbal reports of ownership would be significantly greater for the non-encroached than the encroached hand. Similarly, it was also predicted that encroachment would reduce ownership of the encroached hand in terms of incorporation into the body schema, as measured by pointing, such that reaching would occur as if the real location of the hand was shifted closer to the non-encroached than the encroached fake hand.

METHOD

Participants

A sample of 10 right-handed participants was used (2 male) with a mean age of 21.3 (range 19–24). All were undergraduates or postgraduates at the University of Nottingham, had normal or corrected-to-normal vision, and gave informed consent.

Procedure

Participants were seated at a MIRAGE augmented reality system (University of Nottingham) and given a brief acclimatization period of approximately 10 s, during which they viewed an image of their right hand in its real location moving freely (for details, see Figure 1 and Newport et al., 2010). Participants were asked to position their right hand in a pointing posture (extended index finger), and the experimenter then positioned the hand such that index finger was placed on an unseen start point, where it remained between tasks. In all conditions, participants stroked a sponge-tipped stick with their right index finger at 1 Hz for 20 s. Timing was controlled by a metronome beat for 5 strokes, and then self-maintained, and it was monitored

Figure 1. Real-time video images of the participant's moving hand were displayed with the MIRAGE system, which allows real-time video capture (delay < 17 ms) of the participant's actual limb to be presented from the same position and from the same visual perspective as if viewing the real limb directly. In experimental conditions, participants were presented with two duplicate images of the right hand, both moving synchronously with the real hand. On given trials, a stick appeared to pass through the near, far, or neither hand. A black screen (not shown) was positioned above the forearm to ensure that participants were unable to see any portion of their real limb directly.

by the experimenter referring to a visual analog timer. In the baseline condition, a single image of their right hand was presented, in the same location as the real hand and moving in real time. In the experimental conditions, two images of the right hand were presented, one positioned 6 cm further from the body (far FH) and one 6 cm nearer to the body (near FH) than the actual hand located 18 cm to the right of the body midline. The two images were thus equidistant from the location of the real hand (which could not be seen), did not overlap, and did not cross the body midline. In all conditions, both fake hands moved in synchrony with the unseen real hand. Following 20 s of stroking, the sponge-tipped stick was withdrawn from view but was seen to encroach the far fake hand (FE), the near hand (NE), or neither hand (NeE) on route. Prior to the experimental procedure, participants were informed that they would be shown two right hands but were given no information as to which (if any) corresponded to the position of their real hand. Following stroking and encroachment (or not), participants either verbally responded to questionnaire items (taking approximately 2 min) or pointed three times in succession to a visual target with their right hand (without vision of either real or fake hands). The pointing target was displayed as a red cross against a black background, 10 cm directly in front of the real index finger, and remained on screen for the duration of the reach. At the end of each reach, all participants were able to successfully return to the tactile start point or were repositioned by the experimenter. During the pointing tasks, no hands (real or otherwise) were visible at any point. End-point reach errors were calculated for each condition as the angle measured from the start point between the end location of the movement and the target location (Newport et al., 2010). This was performed for each condition and baseline-corrected by subtracting end-point error measured from the baseline condition for each participant. Questionnaire and pointing responses were made only once for each condition, the order of which was counterbalanced across conditions and participants. Although participants reached three times in each condition, the effects were seen to "wash out" slightly between the first and last reach—probably as a result of active movement updating the felt position of the limb. Although the pattern of results was almost identical between analyses of one and three reaches, only the first reach in each trial is reported here, as it is the location from which the initial reach was planned, before any potential recalibration could take place, that has the greatest theoretical importance.

Questionnaires were presented on-screen with responses given verbally such that the participants were not required to move during this part of the experiment All responses were given according to a 11-point rating scale (0 = strongly disagree, 10 = strongly agree), and the questions were based on those used by Botvinick and Cohen (1998).

Questionnaire A was presented in the baseline conditions only (stroking the sponge-tipped stick with a single unmanipulated image of the real right hand) and consisted of six questions related to participants' experience of their hand as viewed through the screen. This questionnaire was included in order to confirm that participants felt a sense of ownership over a video image of their hand when presented in its real location.

Questionnaire B was presented three times, once for each experimental condition (FE, NE, NeE), and consisted of 12 questions regarding subjects' experience of their limb ownership for both the near and far hand. The aim of questionnaire B was to establish which of the seen limbs were incorporated into participants' body image in each condition.

RESULTS

Questionnaire data

Questionnaire A

The results of questionnaire A, presented in the baseline condition only, demonstrated that participants felt a strong sense of ownership over the hand image (presented in the same location as the real hand) after only 20 s of stroking. All participants gave a score of 8 or greater (out of 10) when presented with the statement, "*It seemed like the image of the hand belonged to me,*" and all gave a score of 7 or greater to the statement, "*It seemed like the image of the hand was my hand.*"

Questionnaire B

Rating scores were collapsed across the questions, *'It seemed like the near/far/both hand(s) belonged to me,' 'It seemed like the near/far/both hand(s) was part of my body,'* and *'It seemed like the touch I felt came from the near/far/both hand(s) touching the sponge,'* to calculate a measure of ownership for each hand in each of the conditions. The mean data (Figure 2) were then entered into a 3 × 3, repeated-measures ANOVA with the factors, Hand Ownership (NEAR, FAR, BOTH) and Hand Encroachment (FE, NE, NeE).

There was a significant main effect identified for Hand Ownership, $F(1, 9) = 6.61, p < .01$, with higher ownership scores recorded when asked about the NEAR (7.89) compared to the FAR (5.22) and BOTH (5.09) hands. A significant main effect was also found for Hand Encroachment, $F(1, 9) = 3.66$, $p < .05$, with higher ownership scores for NeE (7.1) than both NE (5.64) and FE (5.44). There was also a significant interaction, $F(2, 18) = 3.27, p < .05$, the means for which are given in Figure 2. A priori planned comparisons revealed that encroachment of the far hand (FE) significantly reduced ownership scores for the far hand (mean = 3.3) compared to in the non-encroachment (NeN) condition (mean = 6.3), $F(1, 9) = 7.99$, $p < .01$. Encroachment of the near hand (NE), however, made no significant difference to ownership scores for the near hand (mean = 6.5) compared to the non-encroached (NeN) condition (mean = 8.3), $F(1, 9) = 2.87$, $p = .099$.

Pointing data

The baseline-corrected reach error was entered in a one-way ANOVA (FE, NE, and NeE), which yielded a significant result, $F(2, 9) = 5.95$, $p < .05$. A priori planned comparisons revealed no significant difference between NE, −1.22° (12.4), and NeE, 1.54° (11.7), $F(1, 9) = 1.42, p = .249$. However, there was a significant difference between FE and NeE, $F(1, 9) = 4.87$, $p < .05$, such that reach error was greater when the far hand was encroached (FE), mean = 6.66° (12.3), compared to no encroachment (NeE), mean = 1.73°. FE error deviated significantly to the right, consistent with reaches planned with a hand located more to the left, or

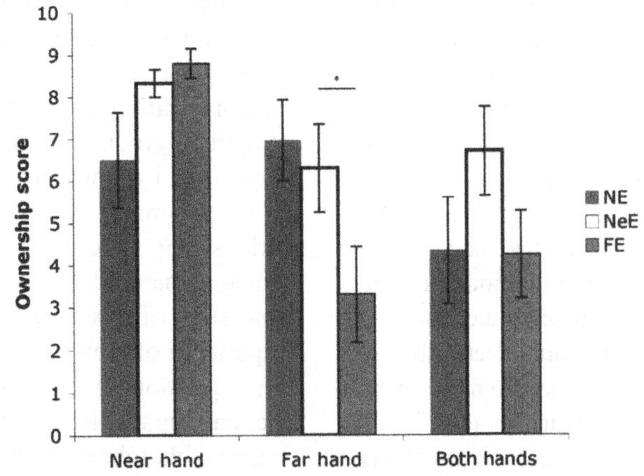

Figure 2. Mean ownership scores for the near and far hands in the near (NE), far (FE), and neither (NeE) encroachment conditions. Ownership was significantly reduced for the far hand in the far encroach condition. Error bars show *SE*.

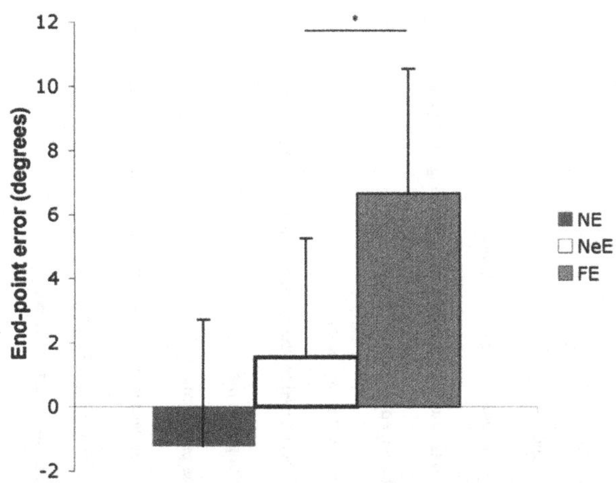

Figure 3. Mean end-point errors (in degrees) for the reaches made with the unseen real hand in the neither encroached (NeE), near encroached (NE), and far encroached (FE) conditions. Negative errors represent a leftward error (toward the body). Errors were significantly rightward when the far hand was encroached, consistent with reaches planned from a location shifted toward the non-encroached near hand. Error bars show *SE*.

closer to the body, than the real hand (for means, see Figure 3) and was equivalent to a shift of approximately 20% of the distance to the near fake hand. Pointing errors did not correlate with ownership scores for either the FE, $r(8) = .2$, $p = .58$, or the NE, $r(8) = .35$, $p = .32$, conditions.

DISCUSSION

The results support the idea that if a fake limb violates our fundamental knowledge of the human body, then this can inhibit incorporation of the limb into representations for the body image and body schema despite synchronous sensory stimulation and action (agency). In this experiment, the encroached hand was disembodied in favor of the synchronous non-encroached hand, but it is interesting to note that this only occurred for the hand that was furthest from the body. When the hand nearest the body was encroached, also violating our knowledge of a real hand, ownership ratings remained high, as participants still reported feeling as though both of the fake hands were part of their body (see Figure 4 for medians and interquartile ranges).

Proximity to the body has been found to affect embodiment of a fake limb into the body schema (Newport et al., 2010) and the body image (Lloyd, 2007, but see Zopf, Savage, & Williams, 2010, for contrasting results in which a distance effect was apparent only for asynchronous stroking). The current experiment expands upon these findings and further demonstrates the importance of closeness to the body. Body parts viewed from the first-person perspective are more likely to be associated with the self and may be processed differently (e.g., Saxe, Jamal, & Powell, 2006). Moreover, the rubber hand illusion is abolished if the rubber hand is rotated to an angle that deviates from the first-person perspective (Tsakiris & Haggard, 2005). It seems intuitive that the internal body template (Tsakiris, 2010) is based on a first-person perspective and that anything deviating from this contradicts the template and so is not considered as part of the body. Distance from the body midline might modulate embodiment in a similar way. It would be very unusual to view another's limb closer to our body midline than our own limb from a first-person perspective (i.e., with the hand facing away from us), and our own limbs, being attached to our body, should be the closest limb appearing to emanate from our body. This may mean that a fake limb nearest the body midline fits the internal body template more closely than a fake or real limb that is more distal. The results from the current study highlight the importance of the relative location to the body midline when attempting to embody or disembody fake limbs within the body image and body schema. This may be important for the design and interpretation of rubber hand studies in which the real hand is often much further from the body than the fake one, although a reviewer might argue that the sample size in the current experiment was relatively modest and that the null effect of encroachment on near-hand ownership ($p = .099$) may benefit from replication in a larger sample.

The results of the pointing data broadly mirror those of the questionnaire data such that when the near hand was encroached, reach trajectories were equivalent to the no-encroachment condition, yet when the far hand was encroached, participants reached as though the position of their actual limb was shifted toward that of the non-encroached near hand. This shift in

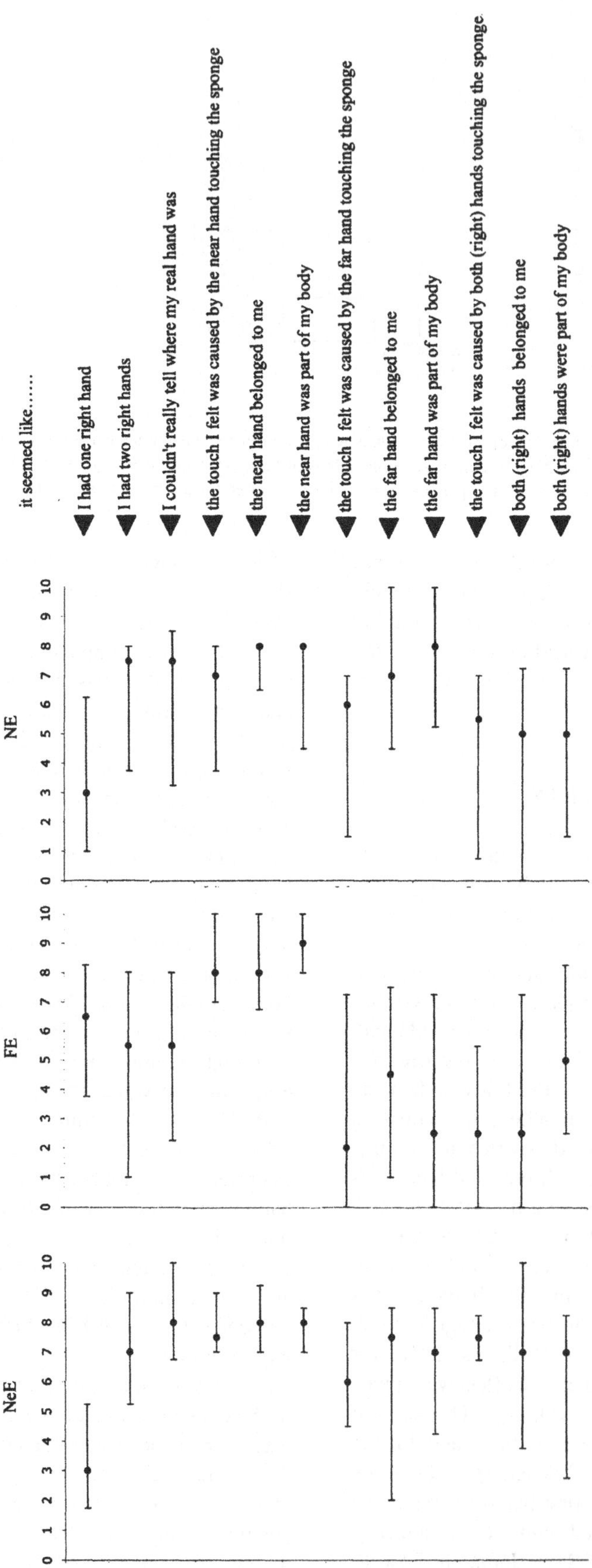

Figure 4. Medians and interquartile ranges for responses to questions of ownership in the neither encroached (NeE), near encroached (NE), and far encroached (FE) conditions. Ownership scores for the far hand were reduced when the far hand was encroached.

trajectory was small compared to the drop in ownership ratings observed from the questionnaire data. Although the direction of effects was similar for image and schema, hand encroachment affected embodiment to different degrees and was not correlated, as is compatible with the idea of separable body representations for action and perception into which fake body parts can be incorporated independently of each other (Preston & Newport, 2011).

The authors are not aware of any previous bodily encroachment experiments in healthy individuals, so discussion of the neural mechanisms involved is rather speculative. There seem to be at least three plausible candidates for the involvement of underlying brain regions: the extrastriate body area (EBA), the insula, and the right temporoparietal junction (rTPJ). Although the specificity of the EBA is still under debate, there is evidence to suggest that it codes for our own body parts as well as for those of other people (Myers & Sowden, 2008). Although all the body parts on view were those of the participant and no distinction between self and other was required in this task, an encroached limb may have violated the brain's body image template and caused a switch from processing the viewed hand as belonging to the participant to belonging to somebody else.

Activity in the insula increases when the body is threatened, when identifying self-actions and is implicated in abnormal representations of the self (Ehrsson, Wiech, Weiskopf, Dolan, & Passingham, 2007; Farrer et al., 2003; Karnath & Baier, 2010). As the encroaching stimulus approached and then passed through personal body space, differential involvement of the insula might be expected, depending on whether the stimulus was perceived as a threat to body space and whether the viewed hand was treated as part of the bodily self before and after encroachment.

Thirdly, the rTPJ has been identified as a potential location for a mechanism that compares visual input with the body image template and is crucial for the incorporation of objects into the body image (for review, see Tsakiris, 2010). In terms of the body schema, however, more dorsal parietal regions may make more likely candidates. In the current experiment, participants were required to reach without vision of the limb, necessitating accurate updating of postural configurations. Recent evidence suggests that maintaining an up-to-date estimate of limb posture for reaching involves the superior parietal lobe supported by connections with premotor areas (Parkinson, Condon, & Jackson, 2010). An alternative view might postulate that the viewed limbs, being repeatedly encroached throughout the experiment, cease to be treated as parts of the body at all and are instead processed as objects, and so do not recruit areas associated with the self and self-action processing. The repeated encroachment may account for the relatively small effect on the body schema.

Of course, it should be noted that having an extra hand is, in itself, a violation of our body template, but it is already well documented that the brain contains multiple representations of the limb (e.g., image and schema) and can quite happily, under certain experimental and neurological conditions, maintain and accept multiple visual representations of the same limb. However, it is important to conduct further experimentation in order to examine the effects of encroachment under the more naturalistic experience of having only one right or left hand.

In summary, encroachment of a fake hand inhibited embodiment, but only when that hand was the furthest from the body. Embodiment of the near hand was more robust, suggesting that this may be closer to the default internal body template representation in the brain (Tsakiris, 2010). The body schema was similarly affected by encroachment, although to a lesser extent, with reaches being planned from a location toward the fake hand, but only when the far hand had been violated. Violation of the near hand had no such effect, again suggesting that the representation of the near hand was the more robust representation. The lack of correlation between the effects of encroachment on measures of body image and body schema supports the growing body of evidence in favor of multiple and dissociable body representations in the brain.

REFERENCES

Armel, K. C., & Ramachandran, V. S. (2003). Projecting sensations to external objects: Evidence from skin conductance response. *Proceedings of Biological Science, 270*, 1499–1506.

Botvinick, M., & Cohen, J. (1998). Rubber hands "feel" touch that eyes see. *Nature, 391*, 756.

Ehrsson, H. H. (2009). How many arms make a pair? Perceptual illusion of having an additional arm. *Perception, 38*, 310–312.

Ehrsson, H. H., Wiech, K., Weiskopf, N., Dolan, R. J., & Passingham, R. E. (2007). Threatening a rubber hand that you feel is yours elicits a cortical anxiety response. *Proceedings of the National Academy of Sciences of the United States of America, 104*, 9828–9833.

Farne, A., Pavani, F., Meneghello, F., & Làdavas, E. (2000). Left tactile extinction following visual stimulation of a rubber hand. *Brain, 123*, 2350–2360.

Farrer, C., Franck, N., Georgieff, N., Frith, C. D., Decety, J., & Jeannerod, M. (2003). Modulating the experience of agency: A positron emission tomography study. *NeuroImage, 18*, 324–333.

Giummarra, M. J., Gibson, S. J., Georgiou-Karistanis, N., & Bradshaw, J. L. (2008). Mechanisms underlying embodiment, disembodiment and loss of embodiment. *Neuroscience and Biobehavioral Reviews*, *32*, 143–160.

Henderson, W. R., & Smyth, G. E. (1948). Phantom limbs. *Journal of Neurology, Neurosurgery, and Psychiatry*, *11* (2), 88–112.

Iriki, A., Tanaka, M., & Iwamura, Y. (1996). Coding of modified body schema during tool use by macaque post-central neurones. *Neuroreport*, *2*, 2325–2330.

Kao, K. L. C., & Goodale, M. A. (2009). Enhanced detection of visual targets. *Neuropsychologia*, *47*, 2454–2463.

Karnath, H. O., & Baier, B. (2010). Right insula for our sense of limb ownership and self-awareness of actions. *Brain Structure and Function*, *214*, 411–417.

Khateb, A., Simon, S. R., Dieguez, S., Lazeyras, F., Momjian-Mayor, I., Blanke, O., et al. (2009). Seeing the phantom: A functional magnetic resonance imaging study of a supernumerary phantom limb. *Annals of Neurology*, *65*, 698–705. doi: 10.1002/ana.21647

Lloyd, D. M. (2007). Spatial limits on referred touch to an alien limb may reflect boundaries of visuo-tactile peripersonal space surrounding the hand. *Brain and Cognition*, *64*, 104–109.

Maravita, A., Spence, C., Kennett, S., & Driver, J. (2002). Tool-use changes multimodal spatial interactions between vision and touch in normal humans. *Cognition*, *83*, 25–34.

Moore, D. G., Goodwin, J. E., George, R., Axelsson, E. L., & Braddick, F. M. B. (2007). Infants perceive human point-light displays as solid forms. *Cognition*, *104*, 377–396.

Myers, A., & Sowden, P. T. (2008). Your hand or mine? The extrastriate body area. *NeuroImage*, *42*, 1669–1677.

Newport, R., Pearce, R., & Preston, C. (2010). Fake hands in action: Embodiment and control of supernumerary limbs. *Experimental Brain Research*, *204*, 385–395.

Parkinson, A., Condon, L., & Jackson, S. R. (2010). Parietal cortex coding of limb posture: In search of the body-schema. *Neuropsychologia*, *48*, 3228–3234.

Pavani, F., Spence, C., & Driver, J. (2000). Visual capture of touch: Out-of-the-body experiences with rubber gloves. *Psychological Science*, *11*, 353–359.

Preston, C., & Newport, R. (2008). Was that me? The right hemisphere and the sense of agency: Embodiment of others and misattribution of self-generated action. In B. Hardy-Vallee & N. Payette (Eds.), *Beyond the brain: Embodied, situated and distributed cognition* (pp. 107–122). Newcastle, UK: Cambridge Scholars Publishing.

Preston, C., & Newport, R. (2011). Evidence for dissociable representations for body image and body schema from a patient with visual neglect. *Neurocase.* Advance online publication. doi: 10.1080/13554794.2010.532504.

Poeck, K. (1964). Phantoms following amputation in early childhood and in congenital absence of limbs. *Cortex*, *1*, 269–275.

Saxe, R., Jamal, N., & Powell, L. (2006). My body or yours? The effect of visual perspective on cortical body representations. *Cerebral Cortex*, *16*, 178–182.

Saxe, R., Tzelnic, T., & Carey, S. (2006). Five-month-old infants know humans are solid, like inanimate objects. *Cognition*, *101*, B1–B8.

Stevens, J. A., Fonlupt, P., Shiffrar, M. A., & Decety, J. (2000). New aspects of motion perception: Selective neural encoding of apparent human movements. *Neuroreport*, *11*, 109–115.

Tsakiris, M., & Haggard, P. (2005). The rubber hand illusion revisited: Visuotactile integration and self-attribution. *Journal of Experimental Psychology: Human Perception and Performance*, *31*, 80–91.

Tsakiris, M. (2010). My body in the brain: A neurocognitive model of body-ownership. *Neuropsychologia*, *48*, 703–712.

Vallar, G., & Ronchi, R. (2009). Somatoparaphrenia: A body delusion. A review of the neuropsychological literature. *Experimental Brain Research*, *192*, 533–551.

Zopf, R., Savage, G., & Williams, M. A. (2010). Crossmodal congruency measures of lateral distance effects on the rubber hand illusion. *Neuropsychologia*, *48*, 713–725.

COGNITIVE NEUROSCIENCE, 2011, 2 (3–4), 171–178

Proprioceptive drift without illusions of ownership for rotated hands in the "rubber hand illusion" paradigm

Henning Holle, Neil McLatchie, Stefanie Maurer, and Jamie Ward

School of Psychology and Sackler Centre for Consciousness Science, University of Sussex, Falmer, Brighton, UK

The rubber hand illusion is one reliable way to experimentally manipulate the experience of body ownership. However, debate continues about the necessary and sufficient conditions eliciting the illusion. We measured proprioceptive drift and the subjective experience (via questionnaire) while manipulating two variables that have been suggested to affect the intensity of the illusion. First, the rubber hand was positioned either in a posturally congruent position, or rotated by 180°. Second, either the anatomically same rubber hand was used, or an anatomically incongruent one. We found in two independent experiments that a rubber hand rotated by 180° leads to increased proprioceptive drift during synchronous visuo-tactile stroking, although it does not lead to feelings of ownership (as measured by questionnaire). This dissociation between drift and ownership suggests that proprioceptive drift is not necessarily a valid proxy for the illusion when using hands rotated by 180°.

Keywords: Body representation; Embodiment; Rotation; Touch; Visuo-tactile.

When we are being touched, the touch sensation has a specific quality which assures us that "my body belongs to me." This feeling of body ownership is typically not experienced when we simply observe other people being touched but is confined to one's own body (but see Holle, Banissy, Wright, Bowling, & Ward, 2011). One way to investigate and experimentally manipulate this experience of body ownership is by means of the rubber hand illusion (RHI) (cf. Tsakiris, Jimenez, & Costantini, 2010). The typical RHI paradigm consists of the following parts: (1) the real hand of the participant, which is concealed from view (e.g., placed in a box); (2) a rubber hand, which is placed before the participant; and (3) tactile stimulation of the real hand and the rubber hand. Under these circumstances, watching the rubber hand being touched in synchrony with one's own unseen hand causes the rubber hand to be attributed to one's own body, and it starts to "feel like it's my hand" (Botvinick & Cohen, 1998). One recent questionnaire devised by Longo, Schuur, Kammers, Tsakiris, and Haggard (2008) identified three components of the RHI: ownership (e.g., "it seemed like the rubber hand was part of my body"), location (e.g., "it seemed like my hand was in the location where the rubber hand was"), and agency (e.g., "it seemed like I could have moved the rubber hand if I had wanted"). In addition to such questionnaire-based data, one can also measure the degree to which the RHI induces a proprioceptive drift by asking participants to report the position of their unseen hand before and after each block. Participants tend to mislocalize their unseen real hand toward the rubber hand while under the illusion (Botvinick & Cohen, 1998), and the amount of this proprioceptive drift is usually correlated with the intensity of the subjective experience (Longo et al., 2008).

Correspondence should be addressed to: Henning Holle, School of Psychology, University of Sussex, Pevensey I Building, Falmer, Brighton, BN1 9QH, UK. E-mail: h.holle@sussex.ac.uk

This study was supported by a grant from the ESRC awarded to J.W. (Grant no. RES-062-23-1150).

http://dx.doi.org/10.1080/17588928.2011.603828

To sum up, the RHI consists of illusory feelings of ownership, location change, and agency. The intensity of the illusion is often correlated with the amount of proprioceptive drift.

One contentious research question has been what the necessary and sufficient conditions to elicit the RHI are. It is rather uncontroversial that synchrony of touch with vision is a necessary condition. Participants tend to report the illusion for synchronous, but not for asynchronous touch. More controversial is the question about what the sufficient conditions are. Armel and Ramachandran (2003) argued for a strong bottom-up view of the RHI. According to their view, synchronous stimulation is both a necessary and sufficient condition to elicit the RHI. In other words, they would predict that synchronous touch is sufficient to incorporate any object into one's own body representation. However, there is now ample evidence that the RHI is not purely bottom-up. For instance, rubber hands but not objects elicit the illusion (Tsakiris & Haggard, 2005). Objects that are closer to the real hand tend to elicit stronger illusions than more distant ones (Lloyd, 2007). Anatomically or posturally incongruent rubber hands tend not to elicit the RHI (Austen, Soto-Faraco, Enns, & Kingston, 2004; Costantini & Haggard, 2007; Ehrsson, Spence, & Passingham, 2004; Pavani, Spence, & Driver, 2000; Tsakiris & Haggard, 2005). In light of such top-down interactions, Tsakiris (2010) has modeled the RHI as an interaction between current multisensory input and internal models of the body. The extracorporeal object first undergoes a goodness-of-fit evaluation, and only extracorporeal objects with a sufficient fit (e.g., an anatomically and posturally matching rubber hand) then undergo an additional comparison, where the observed touch is compared with the felt touch. Tsakiris (2010) stresses that the body model underlying the initial goodness-of-fit evaluation is rather underspecified, and not all aspects of one's own hand, such as skin color (Longo, Schuur, Kammers, Tsakiris, & Haggard, 2009), enter into this comparison.

Given the close relationship between questionnaire measures and proprioceptive drift, one (normally tacit) assumption is that they arise from the same mechanism. Indeed, it is not uncommon for studies to use proprioceptive drift as a proxy of the illusion itself (e.g., Costantini & Haggard, 2007), and to use the degree of drift as an index for the strength of the illusion (Kammers et al., 2009). However, there may be special conditions under which they dissociate. Kammers et al. (2009) found that transcranial magnetic stimulation (TMS) applied over the left inferior parietal lobe disrupted proprioceptive drift measures of the RHI (for right hand stroking), but it did not affect the subjective illusion, as measured via questionnaire. In a rather different paradigm that used only a visual presentation of a rubber hand (presented in a parasagittal mirror), Holmes, Snijders, and Spence (2006) observed a disrupted proprioception of the unseen hand, but this was not accompanied by strong illusory feelings of ownership. The effect was greater for a rubber hand than a wooden block. Holmes and colleagues suggested that visual presentation alone of an appropriate body part (without any tactile stimulus) can result in proprioceptive recalibration.

The present study was motivated by an incidental finding that we subsequently went on to replicate twice (the latter two reported here). Previous studies have failed to demonstrate the RHI for rotations of 180° (Ehrsson et al., 2004), 90° (Tsakiris & Haggard, 2005), or even smaller deviations (Costantini & Haggard, 2007). In two studies, we show that (given a sufficient sample size) proprioceptive drift arising from synchronous visuo-tactile stimulation is present when the anatomically identical hand is rotated by 180° but not when the anatomically mismatching hand is presented at 180° (which can be construed as a mirrored version of the stimulated hand). However, the effect is limited to proprioceptive drift only, and it does not manifest itself in the self-reported illusion (measured via questionnaire).

METHOD

Design

Two separate experiments were conducted. However, the method and results of both studies are presented together because the second experiment was very similar to the first. In both studies, the participants' right hand was always used for tactile stimulation. The first experiment consisted of a 3 × 2 factorial design contrasting rubber hand stimulus (right hand unrotated, right hand rotated, and left hand rotated: R, R180, and L180, respectively) and synchrony (synchronous, asynchronous). The second experiment contained an additional rubber hand stimulus (left hand unrotated), enabling a more sophisticated 2 × 2 × 2 factorial design contrasting hand (left, right), rotation (rotated, unrotated), and synchrony.

Participants

All participants were students at the University of Sussex, who participated either for course credits or for a payment of £5 (corresponding to about 8 US$). They gave informed consent according to the ethics committee of the School of Psychology, University of Sussex. Thirty participants took part in the first experiment (mean age = 23.74; range = 18–50; 7 men), and 39 took part in the second experiment (mean age = 21.1; range = 18–30; 8 men). No participant took part in both. Two participants in Experiment 2 were excluded for having unusually large mean drift scores across all conditions (falling in the upper 5% of the normalized distribution of overall drift scores).

Procedure

Participants were seated at a table, facing the experimenter. On the table was a box, which was open to the sides. Participants were instructed to place their right hand at a marked position inside the box where it was concealed from their view. A rubber hand was also placed at a marked position inside the box, at a position approximately 20 cm to the right of the midline of the participant. The location of the rubber hand was identical in all conditions, differing only in handedness and orientation. The location of the participants' real hand was at an additional 20 cm further to the right. This spatial arrangement of rubber hand and real hand is known to elicit a reliable illusion (Lloyd, 2007). The top cover of the box contained a hole, through which participants could see the rubber hand but not their real hand. Before each trial, the rubber hand was concealed by placing an additional cover on top of the box. A ruler was placed on top of the box, and participants were asked to indicate where they thought that their right index finger was located, by saying the corresponding number on the ruler (the prestimulus position). A different offset was used for the ruler in each trial to reduce memory effects. Next, the ruler and the top cover were removed, and the rubber hand and the real hand were stroked by the experimenter with two identical paintbrushes. Stroking occurred horizontally on the index finger, from knuckle to finger tip and lasted for 1 min. Each stroke lasted approximately 500–1000 ms. The experimenter immediately repositioned the paintbrushes at the knuckle and began the next stroke some 500–1000 ms after the end of the previous stroke. In the asynchronous conditions, asynchrony was randomly varied between 500 and 1000 ms (cf. Tsakiris & Haggard, 2005). For the rotated conditions, the rubber hand was rotated by 180° around the position of the rubber index finger. The direction of stroking in the rotated condition was congruent in hand-centered space (e.g., knuckle to fingertip) but not in external space (see also Tsakiris & Haggard, 2005). Participants were asked to look at the rubber hand throughout the stroking period, which lasted for 60 s. After the stroking had finished, the top cover and ruler were again placed onto the box, and participants were again asked to indicate the position of their right index finger (the poststimulus position). The amount of proprioceptive drift elicited by a trial was subsequently calculated by subtracting the prestimulus position from the poststimulus position.

In Experiment 1, each participant completed a total of 12 trials (two for each of the six conditions) and in Experiment 2, each completed 16 trials (two for each of the eight conditions). The order of the different conditions was randomized for each participant with the constraint that the first and second half of the experiment contained only one of each trial type. Questionnaire data were obtained for the first six trials and eight trials in Experiments 1 and 2, respectively, immediately after the drift measure was taken. We used the short version of the rubber hand questionnaire from Longo et al. (2008). The short version consists of a total of 10 items, referring to three different components of the experience of embodiment during the RHI paradigm: ownership (five items), location (three items), and agency (two items). Participants answered each question on a 7-point Likert scale ranging from −3 ("strongly disagree") to +3 ("strongly agree"), with 0 being "neither agree nor disagree." Questions were randomized separately for each block. Questionnaire data from one participant were incomplete. At the end of each block, participants were asked to remove their hand from the box and wiggle it around to reduce carry-over effects from block to block.

Data analysis

Drift and questionnaire data were analyzed by repeated-measure ANOVAs. Greenhouse–Geisser correction was used where appropriate. In that case, we report the uncorrected degrees of freedom, the corrected p values as well as the correction factor ε. Post-hoc paired t-tests where used where appropriate to evaluate the pattern of significant interactions. We also provide Cohen's d as an effect size measure for these post-hoc tests. Cohen's d gives the standardized difference between two means, and values of 0.2, 0.5, and 0.8 have been suggested to reflect small, medium, and large effects, respectively (Cohen, 1992).

RESULTS

Proprioceptive drift

The results for both experiments are summarized in Figure 1. For Experiment 1, the corresponding 3 × 2 ANOVA for the drift data revealed a significant main effect of Synchrony, $F(1, 24) = 23.45$, $p < .0001$, as well as an interaction between Stimulus Type and Synchrony, $F(2, 48) = 6.21$, $p < .01$, $\varepsilon = .88$. Post-hoc paired *t*-tests (two-tailed) indicated that the simple main effect of Synchrony was significant for conditions R, $t(29) = 5.52$, $p < .0001$, Cohen's $d = 1.29$, and R180, $t(29) = 2.76$, $p < .01$, Cohen's $d = 0.72$, but not for condition L180, $t(29) = 0.61$, $p > .5$. For Experiment 2, the corresponding 2 × 2 × 2 ANOVA revealed main effects of Synchrony, $F(1, 36) = 7.68$, $p < .01$; Hand, $F(1, 36) = 7.24$, $p < .05$; and Rotation, $F(1, 36) = 10.17$, $p < .005$. The interaction between Hand and Rotation was of borderline significance, $F(1, 36) = 3.60$, $p = .066$, but no other interactions approached significance. Post-hoc tests indicated that the simple main effect of Synchrony was significant for conditions R, $t(36) = 2.18$, $p < .05$, Cohen's $d = 1.09$, and R180, $t(36) = 2.04$, $p < .05$, Cohen's $d = 0.44$, but not for conditions L, $t(36) = 1.82$, $p = .08$, and L180, $t(36) = 0.37$, $p > .5$. As such, the main novel finding from these experiments is that one measure of the RHI, proprioceptive drift, can be found for the anatomically congruent hand in at least one orientation other than the standard one (i.e., the anatomically and posturally congruent condition R). The effect is not driven solely by visuo-tactile synchrony because it is not found for the left hand when rotated. It involves the combination of an anatomically congruent, but rotated, body part combined with visuo-tactile synchrony.

One might wonder whether those participants who show the largest proprioceptive drift in the standard condition are the ones who show the greatest effect for the rotated right hand. This was found not to be the case. There was no significant correlation between drift scores between rotated and unrotated right hand (synchronous) across both experiments ($r = .05$, *n.s.*).

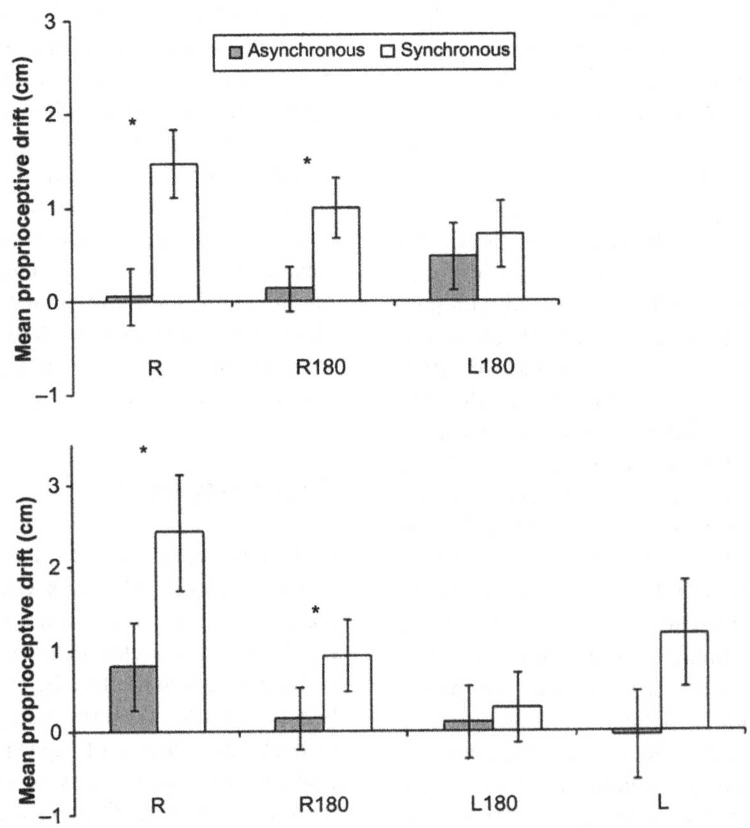

Figure 1. Mean proprioceptive drift for Experiment 1 (top panel) and Experiment 2 (lower panel). Error bars show *SEM*. *$p < .05$.

Questionnaire measures of ownership, location, and agency

For Experiment 1, there were main effects of both Synchrony and Stimulus Type for the ownership subscale: Synchrony, $F(1, 29) = 22.96$, $p < .0001$; Stimulus Type, $F(2, 58) = 12.31$, $p < .001$, $\varepsilon = .87$; the location subscale: Synchrony, $F(1, 29) = 28.90$, $p < .0001$; Stimulus Type, $F(2, 58) = 11.49$, $p < .001$, $\varepsilon = .99$; and the agency subscale: Synchrony, $F(1, 29) = 14.41$, $p < .005$; Stimulus Type, $F(2, 58) = 16.71$, $p < .0005$, $\varepsilon = .70$. In no cases was the interaction significant.[1]

Experiment 2 contained more participants and also enabled the Stimulus Type variable to be broken down into separate Hand and Rotation factors. In this analysis, there were significant Hand × Rotation interactions for all three subscales and these are summarized in Figure 2. For the ownership subscale, there were significant main effects of Synchrony, $F(1, 36) = 62.35$, $p < .001$; Hand, $F(1, 36) = 9.41$, $p < .005$; and Rotation, $F(1, 36) = 50.51$, $p < .001$, together with a significant Hand × Rotation interaction, $F(1, 36) = 17.08$, $p < .001$. No other interactions were significant. As can be seen in Figure 2, the Hand × Rotation interaction seems to be driven by a stronger effect of Rotation for the right hand. A post-hoc paired *t*-test comparing the difference of (right unrotated – right rotated) with (left unrotated – left rotated) confirmed this impression, $t(36) = 4.13$, $p < .001$, Cohen's $d = 0.98$. The same pattern was repeated for the other two subscales. For the location subscale, there were significant main effects of Synchrony, $F(1, 36) = 122.20$, $p < .001$; Hand, $F(1, 36) = 12.46$, $p < .001$; and Rotation, $F(1, 36) = 76.87$, $p < .001$, together with a significant Hand × Rotation interaction, $F(1, 36) = 14.08$, $p < .001$. No other interactions were significant. The Hand × Rotation interaction was again found to be driven by a larger Rotation effect for the right hand, $t(36) = 3.75$, $p < .001$, Cohen's $d = 0.99$. For the agency subscale, there were significant main effects of Synchrony, $F(1, 36) = 26.34$, $p < .001$; Hand, $F(1, 36) = 10.25$, $p < .005$; and Rotation, $F(1, 36) = 24.37$, $p < .001$, together with a significant Hand × Rotation interaction, $F(1, 36) = 24.70$, $p < .001$. Again, the interaction was driven by a larger Rotation effect for the right hand, $t(36) = 4.97$, $p < .001$, Cohen's $d = 0.94$. No other interactions were significant.

Finally, we compared the relationship between questionnaire measures and the proprioceptive drift scores. When we correlated the differences scores (synchronous vs. asynchronous) of the drift data (collapsed across both experiments, $n = 67$) with the respective difference scores of the three questionnaire subscales, we found that drift and questionnaire measures were only correlated in the R condition (ownership: $r = .33$, $p < .01$; location: $r = .29$, $p < .05$; agency: $r = .36$, $p < .01$) but not in the R180 condition (ownership: $r = -.12$, $p > .31$; location: $r = -.18$, $p > .14$; agency: $r = -.07$, $p > .58$). Thus, although there is evidence for the RHI (in terms of proprioceptive drift) in both the R and the R180 condition, only in the former was there a reliable association between drift and subjective measures of the illusion.

DISCUSSION

We conducted two experiments investigating the degree to which the RHI is sensitive to rotations of 180° of either an anatomically congruent or an incongruent hand. Previous research has found that only posturally and anatomically congruent hands are sufficient to elicit the RHI. We found in two independent experiments that a rubber hand rotated by 180° can nonetheless elicit proprioceptive drift (which is often taken as a proxy measure of the illusion) with synchronized visuo-tactile stroking, although this condition does not necessarily lead to an illusory sense of ownership, location, or agency.

Previous studies have sometimes used proprioceptive drift as a proxy for the RHI illusion (Costantini & Haggard, 2007) or used the amount of drift as an index for the strength of the illusion (Kammers et al., 2009). Our data show that this assumption is not always justified. We observed significant drift effects for rotated hands, while at the same time participants denied illusory feelings of ownership. Furthermore, drift and questionnaire scores were found to be correlated only for unrotated, but not for rotated hands, suggesting that the RHI questionnaire and proprioceptive drift are not measuring the same construct, at least when rotated hands are used.

One question our data raise is why previous studies found that rotating the rubber hand abolishes the proprioceptive drift associated with the RHI (Costantini & Haggard, 2007; Tsakiris & Haggard, 2005) whereas we observed that rotations by 180° do not abolish proprioceptive drift. Our results indicate that the drift effect for

[1] Although we found effects of Synchrony on subjective experiences of the illusion, the mean ratings in Experiment 1 tended to be negative (i.e., people tended to choose "disagree" rather than "agree"). If we compare just those people who reported a positive score on the *r*-synchronous ownership questions ($n = 11$) then we still get significant proprioceptive drift effects for the R, $t(10) = 2.88$, $p < .05$, and R180, $t(10) = 2.70$, $p < .05$, conditions but not for the L180 condition, $t(10) = 0.05$, *n.s.*

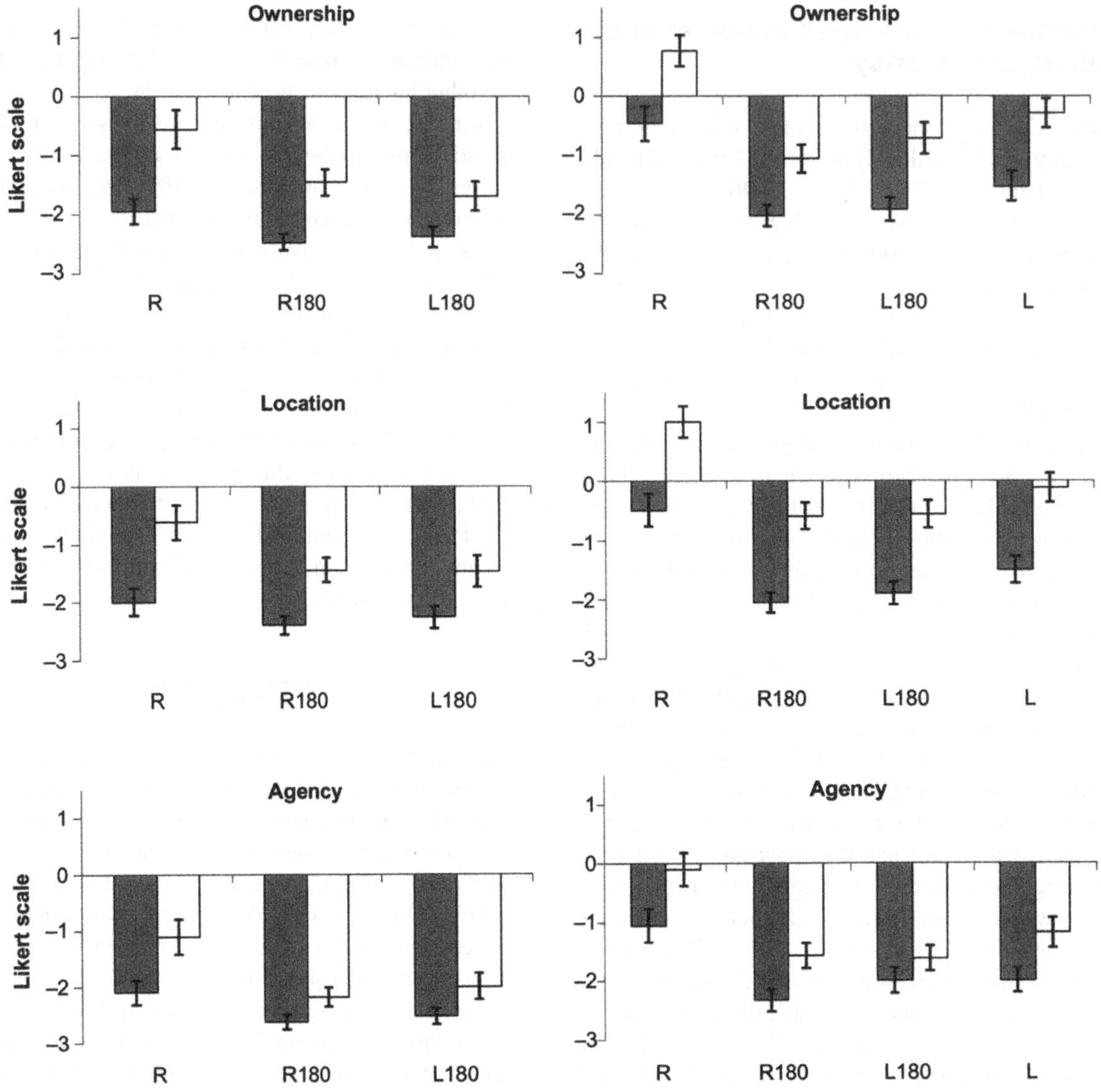

Figure 2. Questionnaire ratings. The Likert scale ranged from −3 ("strongly disagree") to +3 ("strongly agree"), with 0 being "neither agree nor disagree." The left panel shows the data from Experiment 1, and the right panel the data from Experiment 2.

rotated hands is smaller than the effect for unrotated hands and therefore requires larger sample sizes to detect than have been typically used in previous research. We confirmed this impression in a post-hoc power analysis using G-Power (Faul, Erdfelder, Lang, & Buchner 2007), where we found that the two previous studies that measured drift while manipulating postural congruency were indeed underpowered in this respect (Tsakiris & Haggard, 2005: $n = 8$, probability of 42% to detect an effect of Cohen's $d = 0.72$ or greater, 19% when using the smaller effect size obtained in our Experiment 2; Costantini & Haggard, 2007: $n = 11$ in the relevant group, probability of 58% to detect an effect of Cohen's $d = 0.72$ or greater, 26% when using the smaller effect size).

Furthermore, our study underlines the importance of measuring both the subjective experience of the RHI and proprioceptive drift. For example, there were no interactions involving the crucial factor of Synchrony in the questionnaire data (i.e., Synchrony increased feelings of ownership, location, and agency to a similar degree, regardless of what rubber hand was being placed in front of the participant). In contrast, we found interactions involving Synchrony for *some* presentations of the rubber hand in the drift data (compare Figures 1 and 2). Similarly, the presence of the anatomically and posturally correct right hand had an effect in the asynchronous condition as well, not just in the synchronous condition (i.e., Synchrony did not interact with Hand or Rotation in the questionnaire data).

This is consistent with the model of Tsakiris (2010), in which visuo-tactile synchrony and the illusory experience of body ownership are separate processes. However, this model requires refinement to incorporate the data presented in this study. Proprioceptive drift was greater for the anatomically correct hand in the incorrect posture (R180) than an anatomically incorrect hand in the same incorrect posture (L180). This suggests that visuo-tactile synchrony mapped to a hand-centered reference frame in near space may be sufficient to induce reliable proprioceptive drift. However, this does not mean that postural information is unimportant. We only observed a significant correlation between subjective and drift data in the anatomically and posturally correct condition (R). For the posturally incorrect hand (R180), we did observe a significant mislocalization; however, the degree of mislocalization was not associated with increased subjective sensations of ownership, location, and agency. Thus, illusory feelings of ownership may further depend on whether the rubber hand is posturally plausible.

Our finding that 180° rotated body parts can elicit proprioceptive drift in a RHI paradigm fits well with recent neuroimaging studies on observed touch. One emerging theme of this field is that observing a touch event activates the same brain regions that are also active during actual touch, including primary and secondary somatosensory cortex (SI and SII, respectively; for review, see Keysers, Kaas, & Gazzola, 2010). A recent study by Schaefer, Xu, Flor, and Cohen (2009) suggests that although SI and SII are activated during observed touch, they are relatively insensitive to the perspective of observed touch. In that study, similar levels of activity were observed for touch events filmed from an egocentric perspective (as if looking at one's own hands) than for touch events filmed from an allocentric perspective (as if looking at the hands of someone else). This relative insensitivity to the spatial orientation at the initial cortical processing stages may explain why there is a certain tendency to incorporate rubber hands rotated by 180° into one's body representation.

Finally, we also explored in Experiment 2 whether an anatomically incongruent, but posturally congruent rubber hand (condition L) elicits proprioceptive drift or illusory ownership. Unfortunately, the statistical test for the drift effect was not significant ($p = .08$), so we are unable to draw any strong conclusions about this condition. Generally, we observed a great degree of interindividual variation in this condition, with some participants showing strong mislocalizations, whereas others showed no such effect. Further research, especially into what distinguishes participants who report the illusion from those who do not in a rubber hand experiment (interoceptive awareness apparently being one factor; see Tsakiris et al., 2011), may shed further light on this issue.

In conclusion, we report a novel finding that rubber hands rotated by 180° can elicit proprioceptive drift, provided the rubber hand is anatomically congruent to the stimulated real hand. The results suggest that mapping onto a hand-centered space may drive proprioceptive drift toward the rubber hand but that postural alignment is additionally needed for the illusory experience of ownership as measured by questionnaire measures.

REFERENCES

Armel, K. C., & Ramachandran, V. S. (2003). Projecting sensations to external objects: Evidence from skin conductance response. *Proceedings of the Royal Society of London, Series B: Biological Sciences, 270*(1523), 1499–1506.

Austen, E. L., Soto-Faraco, S., Enns, J. T., & Kingstone, A. (2004). Mislocalizations of touch to a fake hand. *Cognitive, Affective, & Behavioral Neuroscience, 4*(2), 170–181.

Botvinick, M., & Cohen, J. (1998). Rubber hands 'feel' touch that eyes see. *Nature, 391*(6669), 756.

Cohen, J. (1992). A power primer. *Psychological Bulletin, 112*(1), 155–159.

Costantini, M., & Haggard, P. (2007). The rubber hand illusion: Sensitivity and reference frame for body ownership. *Consciousness and Cognition, 16*(2), 229–240.

Ehrsson, H. H., Spence, C., & Passingham, R. E. (2004). That's my hand! Activity in premotor cortex reflects feeling of ownership of a limb. *Science, 305*(5685), 875–877.

Faul, F., Erdfelder, E., Lang, A. G., & Buchner, A. (2007). G*Power 3: A flexible statistical power analysis program for the social, behavioral, and biomedical sciences. *Behavior Research Methods, 39*(2), 175–191.

Holle, H., Banissy, M. J., Wright, T., Bowling, N., & Ward, J. (2011). "That's not a real body": Identifying stimulus qualities that modulate synaesthetic experiences of touch. *Consciousness and Cognition, 20*(3), 720–726.

Holmes, N. P., Snijders, H. J., & Spence, C. (2006). Reaching with alien limbs: Visual exposure to prosthetic hands in a mirror biases proprioception without accompanying illusions of ownership. *Perception & Psychophysics, 68*(4), 685–701.

Kammers, M. P., Verhagen, L., Dijkerman, H. C., Hogendoorn, H., De Vignemont, F., & Schutter, D. J. (2009). Is this hand for real? Attenuation of the rubber hand illusion by transcranial magnetic stimulation over the inferior parietal lobule. *Journal of Cognitive Neuroscience, 21*(7), 1311–1320.

Keysers, C., Kaas, J. H., & Gazzola, V. (2010). Somatosensation in social perception. *Nature Reviews Neuroscience, 11*(6), 417–428.

Lloyd, D. M. (2007). Spatial limits on referred touch to an alien limb may reflect boundaries of visuo-tactile peripersonal space surrounding the hand. *Brain and Cognition, 64*(1), 104–109.

Longo, M. R., Schuur, F., Kammers, M. P., Tsakiris, M., & Haggard, P. (2008). What is embodiment? A psychometric approach. *Cognition, 107*(3), 978–998.

Longo, M. R., Schuur, F., Kammers, M. P., Tsakiris, M., & Haggard, P. (2009). Self awareness and the body image. *Acta Psychologica, 132*(2), 166–172.

Pavani, F., Spence, C., & Driver, J. (2000). Visual capture of touch: Out-of-the-body experiences with rubber gloves. *Psychological Science, 11*(5), 353–359.

Schaefer, M., Xu, B., Flor, H., & Cohen, L. G. (2009). Effects of different viewing perspectives on somatosensory activations during observation of touch. *Human Brain Mapping, 30*(9), 2722–2730.

Tsakiris, M. (2010). My body in the brain: A neurocognitive model of body-ownership. *Neuropsychologia, 48*(3), 703–712.

Tsakiris, M., Carpenter, L., James, D., & Fotopoulou, A. (2010). Hands only illusion: Multisensory integration elicits sense of ownership for body parts but not for non-corporeal objects. *Experimental Brain Research, 204*(3), 343–352.

Tsakiris, M., & Haggard, P. (2005). The rubber hand illusion revisited: Visuotactile integration and self-attribution. *Journal of Experimental Psychology: Human Perception and Performance, 31*(1), 80–91.

Tsakiris, M., Jimenez, A. T., & Costantini, M. (2011). Just a heartbeat away from one's body: Interoceptive sensitivity predicts malleability of body-representations. *Proceedings of the Royal Society of London, Series B: Biological Sciences*. Advance online publication. doi: 10.1098/rspb.2010.2547.

Disownership and disembodiment of the real limb without visuoproprioceptive mismatch

Roger Newport[1] and Catherine Preston[2]

[1]School of Psychology, University of Nottingham, Nottingham, UK
[2]Division of Psychology, Nottingham Trent University, Nottingham, UK

The brain is able to sustain multiple bodily representations, and it is still unclear whether real body parts can be disembodied during the embodiment of fake or illusory body parts. The current experiment attempted to disembody the real hand without the visuoproprioceptive mismatch thought to compromise disembodiment in the rubber hand illusion by providing asynchronous feedback of the hand viewed in its actual location. Participants disowned their real hand and claimed ownership of a synchronous fake hand as measured by ownership questionnaires and skin conductance responses to perceived threat. In contrast, the body schema, as indexed by manual pointing, was only mildly affected, further supporting the distinction between bodily representations. Conflicts in time were more disruptive than conflicts in space for body ownership, while the effect was less so for body schema, schema being more affected by the proximity of vision and proprioception.

Keywords: Body image; Body schema; Sensorimotor integration; MIRAGE; Disembodiment; Multisensory synchrony.

It has been well documented that fake limbs can be incorporated into the body image. The rubber hand illusion (RHI) is a simple demonstration in which synchronous stroking of an unseen real hand and a seen rubber hand results in feelings of ownership over the rubber hand, as indicated by responses given to ownership questionnaires and perceived shifts in the position of the real hand toward that of the rubber hand (e.g., Botvinick & Cohen, 1998). Although it has been established that embodiment of a fake limb occurs during the RHI, it remains unclear what happens to the actual limb; whether it is disembodied and replaced by the rubber hand or whether representations of both hands exist in parallel.

That the brain can support multiple representations of the limb seems clear. For example, researchers have recorded dissociation for embodiment of the rubber hand into body image and body schema, supporting a distinction between limb representations. Kammers, de Vignemont, Verhagen, and Dijkerman (2009a) measured the effect of the RHI on reaching movements, finding that, although participants reported a perceived shift in hand location following synchronous stroking of the real and rubber hands, when they used the hand for action their reaches were unaffected by the illusion. This demonstrated that while the RHI can lead to perceptual embodiment of a fake limb, it was the representation of the real limb that was used for action (i.e., the limb had been incorporated into the body image, but not the body schema). The dissociation of perceptual and motor embodiment implies that multiple representations of the same limb can be maintained simultaneously, something that has also been demonstrated with multiple fake limbs (Ehrsson, 2009; Newport, Pearce, & Preston, 2010). Newport et al. demonstrated that synchronous active illusory hands can penetrate both the body schema and the body image: When presented with two identical and synchronously moving hands, both could be incorporated into the body image while only one was incorporated into the body schema to be used for action. However, this manipulation did not

Correspondence should be addressed to: Roger Newport, School of Psychology, University of Nottingham, Nottingham NG7 2RD, UK. E-mail: roger.newport@nottingham.ac.uk

http://dx.doi.org/10.1080/17588928.2011.565120

provide direct evidence for disembodiment of the real hand because although the hand used for pointing could be switched from one illusory hand to the other, neither hand was in the same physical location as the real hand. Being able to maintain multiple representations may mean that in ambiguous situations such as the RHI, the real hand is not, and does not need to be, fully disembodied, and it is yet to be determined whether the real hand *can* be disembodied in the absence of neurological abnormality, a question that needs to be addressed before we can fully understand how the body is represented in the brain.

Evidence for disembodiment of the real limb during RHI is mixed. Moseley et al. (2008) recorded a drop in skin temperature for the real limb when participants claimed ownership of a fake limb during a synchronous RHI procedure, and this has been interpreted as an indicator of disembodiment of the real limb. However, Folegatti, de Vignemont, Pavani, Rossetti, and Farnè (2009) argue against disembodiment. They induced a mismatch between the seen and felt positions of the real hand, using prism goggles, and observed a slowing of reaction times in response to tactile stimuli that was comparable to those observed following synchronous RHI stimulation. Because the seen and felt positions of the hand are also different in the RHI, Folegatti et al. suggested that somatosensory changes observed in the real hand during the RHI might be explained by visuoproprioceptive conflict rather than disownership of the real hand.

Using a paradigm that has previously demonstrated embodiment of fake limbs into both body image and body schema, the current experiment was designed to investigate whether the hand could be disembodied from body image and/or schema without the visuoproprioceptive mismatch inherent in RHI procedures. A MIRAGE augmented reality system (University of Nottingham) (e.g., Newport et al., 2010) was used to display real-time images of both the real and a simultaneously presented duplicate hand while actively stroking a toothbrush. Of the two representations of the participant's own hand, one appeared in the same physical location as the real hand while the other was offset slightly to the left or right of the real hand. Temporal delays could be applied independently to either hand in order to produce hand-specific asynchronous feedback such that the real or fake hand could be temporally synchronous with reality ("real" here refers to the image of the hand in the same location as the actual hand). Participants responded to questions of ownership and pointed to targets without vision of the hand after a short period of illusion induction. In a second experiment, skin conductance responses (SCR) were employed as a more implicit measurement of ownership (Armel & Ramachandran, 2003) in response to "attacks" toward the hands by virtual threatening stimuli.

The primary comparisons of interest here are those between situations in which the hand seen in the same location as the actual hand (the "real" hand) moves asynchronously while the hand seen in an alternative location (the "fake" hand) moves synchronously compared to when the fake hand is asynchronous, but the real hand is synchronous. If it is possible to disown the real hand in its real location by simultaneously providing a synchronous fake hand in an alternative location, then responses to the synchronous fake hand should be favored in comparison to the asynchronous real hand. It is equally possible, however, that the brain can maintain representations for any number of limbs and that the representation for the real limb is not discarded, but exists in parallel, perhaps in a reduced form. If so, SCR should be similar for both the old and the newly acquired limb locations while pointing responses might be initiated as though from the real hand location.

METHOD

Two experiments were conducted, but as the procedures for both were very similar they shall be presented together.

Participants

Ten healthy volunteers (3 male, 3 left-handed) with a mean age of 29 years (range 24–45) participated in Experiment 1, and 10 different participants (7 female, all right-handed) with a mean age of 22.9 years (ranging from 19 to 27) took part in Experiment 2. All were undergraduates or postgraduates at the University of Nottingham, UK, had normal or corrected to normal vision, were naive as to the purpose of the experiment, and had not participated in previous illusory limb studies. Both procedures were approved by the School of Psychology ethics committee.

Materials

Real-time (delay ~17 ms) video images of the participant's actual (left) hand were displayed, using the MIRAGE augmented reality system (see Newport et al., 2010 and Figure 1a). For this experiment, manipulation involved duplicating the hand image so that while one image always represented the hand in its real location (real hand), the duplicated image (fake hand) was presented 6 cm either to the left or to the right of the actual hand (measurements relative to the center of each index finger). Participants were informed that they may be shown an extra hand but were given no information as to which (if any) was their real hand. The left hand was used because most clinical manifestations of supernumerary limbs involve the left side of the body.

Procedure

Prior to the experiment, participants were given a brief acclimatization period (10 s) during which they viewed an unmanipulated image of their left hand moving freely (i.e., in real time and in its real location). For each experimental trial, the participant's index finger was placed on one of two unseen tactile start locations pseudorandomized throughout the experiment. In all conditions, participants stroked the bristles of a toothbrush at 1 Hz for 20 s with their left index finger while viewing one or two images of their left hand. In the baseline condition, only the real hand was shown. In the experimental conditions, both the real and fake hands were shown with a 0.5-s delay applied to either the fake or real hand, so that the stroking was synchronous for the real or fake hand respectively. "Real" hand refers to the hand viewed in the same location as the actual hand; "fake" refers to the offset hand.

Experiment 1

After each period of stroking, participants either verbally responded to questionnaire items or pointed to visual targets with their left hand without vision of the hand. Vision of the hand was removed ~3–5 s before the appearance of the first target and remained occluded throughout the pointing block. The pointing task consisted of three reaches toward one of three targets that was displayed as a red cross against a black background 6 cm forward of the index finger and either 3 cm to the right of the rightmost hand position, 3 cm to the left of the leftmost hand position, or between the two hands (i.e., 3 cm to the left and right, respectively; see Figure 1c). Handpaths were recorded by tracking a 5-mm, infrared reflective marker attached to the nail of the index finger of the reaching hand, using an in-house system recording at 60 Hz (resolution 1.65 mm). There were eight combinations of actual hand position (2), fake hand relative to real hand (2), and real/fake hand synchrony (2), with nine pointing trials and one questionnaire for each combination. The order of presentation of questionnaires and reaching was counterbalanced across and within participants.

Two 10-point rating scale questionnaires were presented via MIRAGE (loosely based on Botvinick & Cohen, 1998), with participants responding verbally so that the hand could not be viewed and was not required to move. A single-hand baseline questionnaire included six questions related to participants' experience of their hand as viewed through the screen, and it was presented only after the baseline condition to confirm that participants felt a sense of ownership over a single video-image of their real hand. A multiple-hands questionnaire, presented in all the other conditions, included 14 questions regarding the experience of ownership for the left, right, and both hand images.

Experiment 2

The procedure was the same as that for Experiment 1 except that, instead of being required to point to targets, participants saw the experimenter's hand, holding a sharp implement, suddenly thrust toward the index finger of the fake or real hand. Threatening stimuli were produced by prerecording images of a hand holding a sharp stick, a needle and syringe, a fork, or a pair of scissors (Figure 1b). A Biopac MP150 data acquisition system (BIOPAC Systems, Inc., Goleta, California) measured skin conductance from sensors on the index and middle fingers of the right hand, and each combination of hand, synchrony, and location was attacked three times in total. A post-experimental questionnaire showed that participants rated the visual stimuli as mildly threatening ($M = 5.46$, out of 10; $SD = 2.89$) and mildly realistic ($M = 5.8$, $SD = 3.48$).

RESULTS

Single-hand questionnaire

Viewing the image of the real hand elicited a strong sense of ownership. After only 20-s stimulation, participants were able to treat the video image of their left hand as if it were their actual left hand: when rating the strength of feeling on a numeric rating scale from 0 to 10 (0 = not at all, 10 = completely), 8 out of 10 participants gave a score of 10 to the statement, "It seemed like the image of the hand was my hand" and 9 out of 10 gave a score of 10 to the statement, "It seemed like the image of the hand belonged to me."

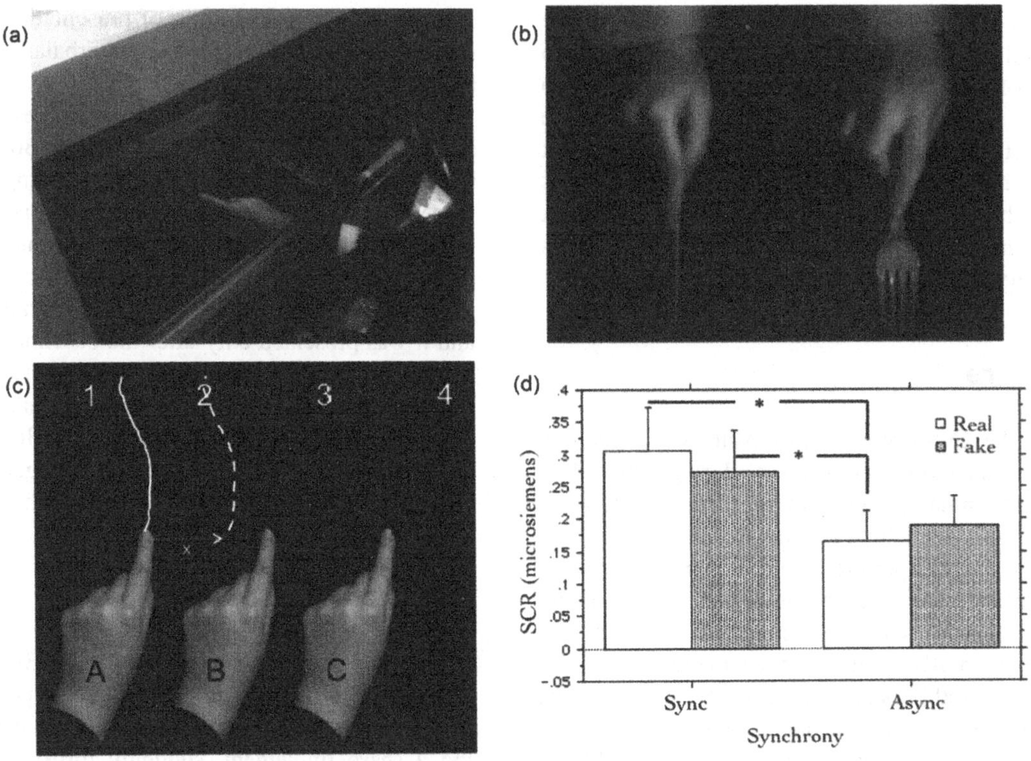

Figure 1. (a) Participant's view of a single hand via MIRAGE. The location and angles of the camera and mirror in MIRAGE are such that images of the participant's limb appear in the same position and from the same perspective as if viewing the limb directly. Captured images can be manipulated online by in-house software with a delay of ~17 ms. A black fabric bib (not shown) ensured that participants were unable to see any portion of their limb outside MIRAGE. (b) Examples of threatening visual stimuli used in the experiment. The images were life-size and moved toward the index finger of either the near or far hand with increasing velocity (bell-shaped velocity profile), reversing when within 5 mm of the finger while at a velocity of 450 mm/s. (c) Schematic representation (not to scale) showing the possible locations of the fake and real hands (A, B, C) and pointing targets (1, 2, 3, 4). Targets 1–3 could be shown for fake/real hand combinations A and B and targets 2–4 could be shown for fake/real hand combinations B and C. Movement origin shifts (x) were calculated by translating the actual reach (solid line) so that it passed through the target. In this example, the reach was made by the real hand in location A toward target 2, with the fake hand located at B. Translating the reach to pass through target 2 shows that this reach was initiated as though the hand were shifted toward the fake hand at location B. (d) Mean skin conductance in response to visual threats in each condition, with standard error bars. Asterisks indicate $p < .05$.

Multiple-hands questionnaire

For statistical comparison of the sense of ownership for each hand, scores from the items "It seemed like the left/right/both hands belonged to me" and "It seemed like the left/right/both hand was part of my body" were collapsed such that a measure of ownership of the real, fake, and both hands was computed for each condition. Overall, the strength of ownership feeling for the synchronous hand (M = 8.74, SD = 1.09) was higher than that of the asynchronous hand (M = 3.21, SD = 2.51) and having two left hands (M = 3.53, SD = 2.36). Planned comparisons were conducted to investigate the difference in the sense of ownership for the real hand when it was asynchronous (fake hand synchronous) compared to when it was synchronous (fake hand asynchronous). This difference was significant, $F(1, 9) = 112.34$, $p < .001$, with ownership scores being significantly lower for the real asynchronous hand (M = 2.83, SD = 2.73) than the real synchronous hand (M = 9.21, SD = 1.31). A significant effect was also found between real asynchronous and fake synchronous, $F(1, 9) = 77.53$, $p < .001$, with sense of ownership scores significantly lower for the real asynchronous hand (M = 2.83, SD = 2.73) than the fake synchronous hand (M = 8.26, SD = 1.77). Medians and interquartile ranges are given in Figure 2.

Pointing data

The start and end locations of each reach were determined by the velocity of the finger rising above and falling below 50 mm/s respectively. To assess the influence of synchrony on reaching, handpath endpoints were aligned with the target location so that the handpaths could be backward plotted to determine the

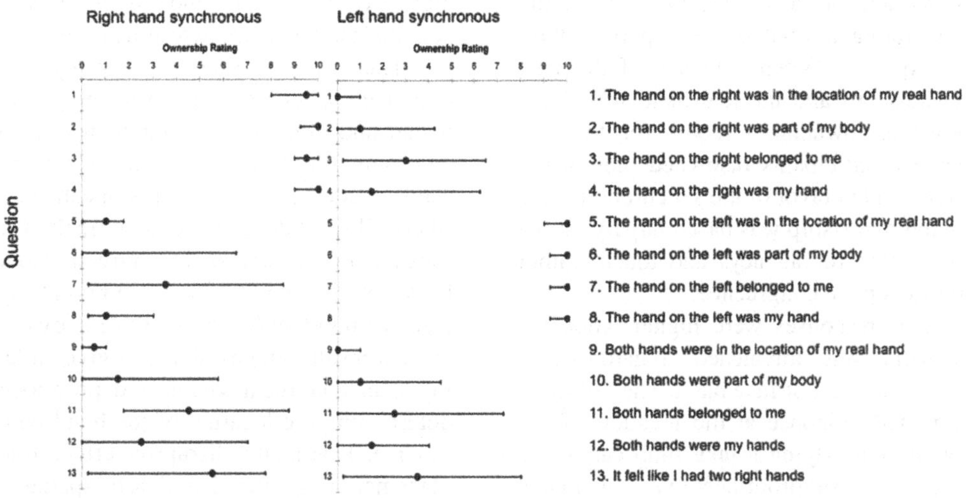

Figure 2. Median and interquartile ranges for ownership ratings given when the hand on either the right or the left was synchronous. 0 reflects low ownership; 10 reflects high ownership.

origin of the movement vector. Reaches were recorded from the position of the actual hand, rather than the image of the real or fake hands; thus, reaches planned from the location of the image of the real hand should originate from the real hand whereas reaches planned with the representation of the fake hand should induce a start location error to the left or right, depending on its location relative to the real hand (see calculation of x in Figure 1c). In order to facilitate analysis, errors were calculated as positive if the induced start location was indicative of a shift toward pointing with the fake hand, regardless of whether the fake hand was to the left or right of the real hand. A t-test was performed between the mean error scores for the conditions in which the real hand was synchronous (fake hand asynchronous) and when the real hand was asynchronous (fake hand synchronous) and this test revealed a significant difference, $t(9) = -6.52, p < .001$. While reaches carried out when the real hand was synchronous were planned from close to the real hand ($M = 4.8$ mm, $SD = 8.5$), when the real hand was asynchronous, reaches deviated as though planned from a location toward that of the fake hand ($M = 26.7$ mm, $SD = 8.5$). This represents a shift toward the fake hand of around 40% of the total distance.

Skin conductance data

Peak and greatest single rise in skin conductance were calculated for the 4-s period immediately following the introduction of a visually threatening stimulus. Both sets of results showed identical patterns, so only the peak data will be reported here for brevity. Planned comparisons revealed that the response to threat was greater for the real hand when synchronous than when asynchronous, $F(1, 8) = 9.93, p < .05$, and also that the response for the fake hand when synchronous was greater than for the real hand when asynchronous, $F(1, 8) = 5.82, p < .05$ (see Figure 1d for means). Taken together, these results suggest that the real hand, when asynchronous, was strongly disembodied in favor of the synchronous fake hand. The pattern of results matched those for subjective limb ownership in Experiment 1, suggesting that they may both be underpinned by the same neural mechanisms, an idea supported by imaging and patient data identifying a role for the insula in both (Ehrsson, Wiech, Weiskopf, Dolan, & Passingham, 2007; Karnath & Baier, 2010).

GENERAL DISCUSSION

The results across the two experiments support a distinction between the perceptual and motor aspects of embodiment. The real hand was disowned equally for both subjective (questionnaire) and objective (SCR) measurements of ownership while the sense of limb location required for action proved more robust and was only partially relocated away from the perceptually disembodied hand toward the embodied hand. As with the traditional RHI, when visual feedback was synchronous with reality for the fake hand, participants claimed ownership of the fake hand, giving significantly greater ownership scores than when the fake hand was asynchronous. When the real hand was synchronous, participants also gave high ownership scores, demonstrating normal ownership over their real hand, as expected. When feedback of the fake hand was synchronous and the real hand was asynchronous, however, ownership scores for the real hand dropped, as participants disowned their asynchronous actual hand in favor of the synchronous fake hand. Thus, hand ownership was based largely on the congruence of action (or agency) and touch, rather than visuoproprioceptive congruence.

Similarly, SCR responses were higher when the synchronous hand was threatened, regardless of whether it was for the real or fake hand. The failure to elicit a strong SCR response at the location of the actual limb while embodying a fake limb cannot be explained by a visuoproprioceptive mismatch between the position of the actual limb and the seen and felt position of the hand, as they were in the same location. The coupling between ownership and the response to threat to the body adds support to the notion that these are mediated by similar neural networks. Neuropsychological and imaging evidence points to a prominent role of the insula in the sense of self and the phenomenal experience of limb ownership. Patients with disorders of limb ownership tend to have lesions to the right insula (reviewed by Karnath & Baier, 2010); the proprioceptive drift associated with limb ownership in the RHI activates the right insula (Tsakiris, Hesse, Boy, Haggard, & Fink, 2007), and activity in the left insula correlates with perceived limb ownership under threatening conditions (Ehrsson et al., 2007).

For the pointing data, although participants reached with the real hand when the real hand was synchronous, the shift toward the synchronous fake hand, while significant, was far from complete at 40%. In this case, it seems that the representation of the limb used for action had not been discarded, but relocated slightly toward the location of the fake hand. The shift is not as substantial as that found in a similar recent experiment in which both subjective ownership and a strong shift in the location of the limb as used by the body schema were demonstrated within the same paradigm (Newport et al., 2010). However, in that case, none of the fake limbs to be embodied were in the same location as the actual limb, perhaps forcing the central nervous system (CNS) to make the best use of the incorrect visual information available. In the current experiment, there was always a visual representation of the limb in the same location as the proprioceptively informed position of the actual limb. Given the close proximity of visual and proprioceptive information in this case, and the capacity for congruent gaze to boost proprioceptive signals of the unseen hand (Newport, Hindle, & Jackson, 2001), the incomplete shift is perhaps not surprising. The parietal lobes of the monkey contain neurons that seem specifically tuned to integrating congruent visual and other sensory input in order to construct and maintain the body schema (Graziano, Cooke, & Taylor, 2000), and it is this mechanism that most likely allowed the CNS to keep track of the true limb location in the face of incongruent movement feedback. While multisensory temporal synchrony is clearly important for the sense of body ownership, the current results suggest that it is given a lesser weighting than congruent vision and proprioception when determining the location of the hand within the body schema. That is, the disruptive effect of temporal displacement is greater than that of spatial displacement for ownership, while the body schema is more heavily influenced by spatial congruence. In RHI studies for which the body schema is resistant to the illusion, the rubber hand can be placed as much as 15–20 cm away from the real hand. At this distance, mechanisms constructing the body schema can discount visual information as erroneous and more heavily weight proprioception and other cues in order to determine the location of the limb for action.

The effect on body schema is different from those observed in some previous studies and may be due to methodological differences such as employing bimanual reaches that require processing of the unmanipulated limb (Kammers et al., 2009a) or unimanual reaches following manipulation of a physically and temporally displaced limb (Kammers, Longo, Tsakiris, Dijkerman, & Haggard, 2009b). The latter study employed a displaced video representation of the hand that was also temporally delayed by 100 ms. Such small delays can cause movements to be judged as no longer completely self-generated (Farrer, Bouchereau, Jeanerrod, & Franck, 2008), and as the current experiment shows, while the influence of spatial congruence may be greater, temporal asynchrony still considerably influences incorporation into the body schema.

In summary, these findings support the idea that the real hand can be disembodied in healthy controls in terms of ownership, but not so much for the limb within the body schema, further supporting a distinction between image and schema body representations. While participants had a strong sense of ownership over a fake hand, while at the same time disowning their real hand, low-level motor control was only partially influenced, and the location of the real hand was not completely replaced by that of the fake hand. It is more likely that the start location of the reach was computed from a weighting of the available visual and proprioceptive information, and, due to the close congruence of these, hand position was not completely overridden by the high-level sense of limb ownership or movement agency.

REFERENCES

Armel, K. C., & Ramachandran, V. S. (2003). Projecting sensations to external objects: Evidence from skin conductance response. *Proceedings of Biological Science, 270*, 1499–1506.

Botvinick, M., & Cohen, J. (1998). Rubber hands "feel" touch that eyes see. *Nature, 391*, 19.

Ehrsson, H. H. (2009). How many arms make a pair? Perceptual illusion of having an additional arm. *Perception, 38*, 310–312.

Ehrsson, H. H., Spence, C., & Passingham, R. E. (2004). That's my hand! Activity in premotor cortex reflects feeling of ownership of a limb. *Science, 305*, 875–877.

Ehrsson, H. H., Wiech, K., Weiskopf, N., Dolan, R. J., & Passingham, R. E. (2007). Threatening a rubber hand that you feel is yours elicits a cortical anxiety response. *Proceedings of the National Academy of Sciences of the United States of America, 104*, 9828–9833.

Farrer, C., Bouchereau, M., Jeanerrod, M., & Franck, N. (2008). Effect of distorted visual feedback on the sense of agency. *Behavioural Neurology, 19*(1–2), 53–57.

Folegatti, A., de Vignemont, F., Pavani, F., Rossetti, Y., & Farnè, A. (2009). Losing one's hand: Visual-proprioceptive conflict affects touch perception. *PLoS One, 4*(9), e6920.

Graziano, M. S., Cooke, D. F., & Taylor, C. S. (2000). Coding the location of the arm by sight. *Science, 290*(5497), 1782–1786.

Kammers, M. P., de Vignemont, F., Verhagen, L., & Dijkerman, H. C. (2009a). The rubber hand illusion in action. *Neuropsychologia, 47*, 204–211.

Kammers, M. P. M., Longo, M. R., Tsakiris, M., Dijkerman, C. H., & Haggard, P. (2009b). Specificity and coherence of body representations. *Perception, 38*, 1804–1820.

Karnath, H.-O., & Baier, B. (2010). Right insula for our sense of limb ownership and self-awareness of actions. *Brain Structure and Function, 214*, 411–417.

Moseley, G. L., Olthof, N., Venema, A., Don, S., Wijers, M., Gallace, A., et al. (2008). Psychologically induced cooling of a specific body part caused by the illusory ownership of an artificial counterpart. *Proceedings of the National Academy of Sciences of the United States of America, 105*(35), 13169–13173.

Newport, R., Hindle, J. V., & Jackson, S. R. (2001). Links between vision and somatosensation: Vision can improve the felt position of the unseen hand. *Current Biology, 11*(12), 975–980.

Newport, R., Pearce, R., & Preston, C. (2010). Fake hands in action: Embodiment and control of supernumerary limbs. *Experimental Brain Research, 204*, 385–395.

Tsakiris, M., Hesse, M. D., Boy, C., Haggard, P., Fink, G. R. (2007). Neural signatures of body ownership: A sensory network for bodily self-consciousness. *Cerebral Cortex, 17*, 2235–2244.

COGNITIVE NEUROSCIENCE, 2011, 2 (3–4), 186–226 (includes Discussion Paper, Commentaries, and Reply)

Discussion Paper

The role of occipitotemporal body-selective regions in person perception

Paul E. Downing[1] and Marius V. Peelen[2]

[1]Wales Institute of Cognitive Neuroscience, School of Psychology, Bangor University, Bangor, UK
[2]Center for Mind/Brain Sciences, University of Trento, Rovereto, Italy

The visual appearance of others' bodies is a powerful source of information about the people around us. This information is implicit in the stimulus and must be extracted and made explicit by the coordination of activity in multiple cortical areas. Here we consider the contribution to this process of two strongly body-selective occipitotemporal regions identified in human neuroimaging experiments: the extrastriate body area (EBA) and the fusiform body area (FBA). We address the evidence and arguments behind numerous recent proposals that EBA and FBA build explicit representations of identity, emotion, body movements, or goal-directed actions from the visual appearance of bodies, and also explore the contribution of these regions to motor control. We argue that the current evidence does not support a model in which EBA and FBA directly perform any of these higher-level functions. Instead, we argue that these regions comprise populations of neurons that encode fine details of the shape and posture of the bodies of people in the current percept. In doing so, they provide a powerful but cognitively unelaborated perceptual framework that allows other cortical systems to exploit the rich, socially relevant information that is conveyed by the body.

Keywords: Body perception; Extrastriate body area; Fusiform body area; Action perception; Emotional gestures; Self–other distinction.

The appearance of the human body provides important visual signals. We see some of our own movements, and to an extent these are guided visually. Visual body information also provides myriad social signals about other people—who they are, what they are doing, how they are feeling, and the like. Accordingly, the brain devotes considerable cortical resources to representing visual information about the human body (Berlucchi & Aglioti, 2010; Minnebusch & Daum, 2009; Peelen & Downing, 2007a; Schwoebel & Coslett, 2005). Clear evidence for specialized visual representations of the body—and a link between these and the temporal lobes—extends well into the previous century, and can be found in neuropsychology (Konorski, 1967) and in some of the earliest work on the visual response properties of single units in the macaque temporal cortex (Gross, 1992).

Many of the more recent studies of the neural underpinnings of body perception in the human temporal lobes have converged around two strongly body-selective, focal regions of the human brain, identified with functional magnetic resonance imaging (fMRI) and confirmed with other methods. These are the extrastriate body area (EBA) (Downing, Jiang, Shuman, & Kanwisher, 2001) found in the posterior inferior temporal sulcus/middle temporal gyrus, and the fusiform body area (FBA) (Peelen & Downing, 2005a; Schwarzlose, Baker, & Kanwisher, 2005) found ventrally on the fusiform gyrus (Figure 1). The discovery of these regions has triggered a large volume and variety

Correspondence should be addressed to: Paul E. Downing, Wales Institute of Cognitive Neuroscience, School of Psychology, Bangor University, Bangor, Gwynedd LL57 2AS, UK. Email: p.downing@bangor.ac.uk

http://dx.doi.org/10.1080/17588928.2011.582945

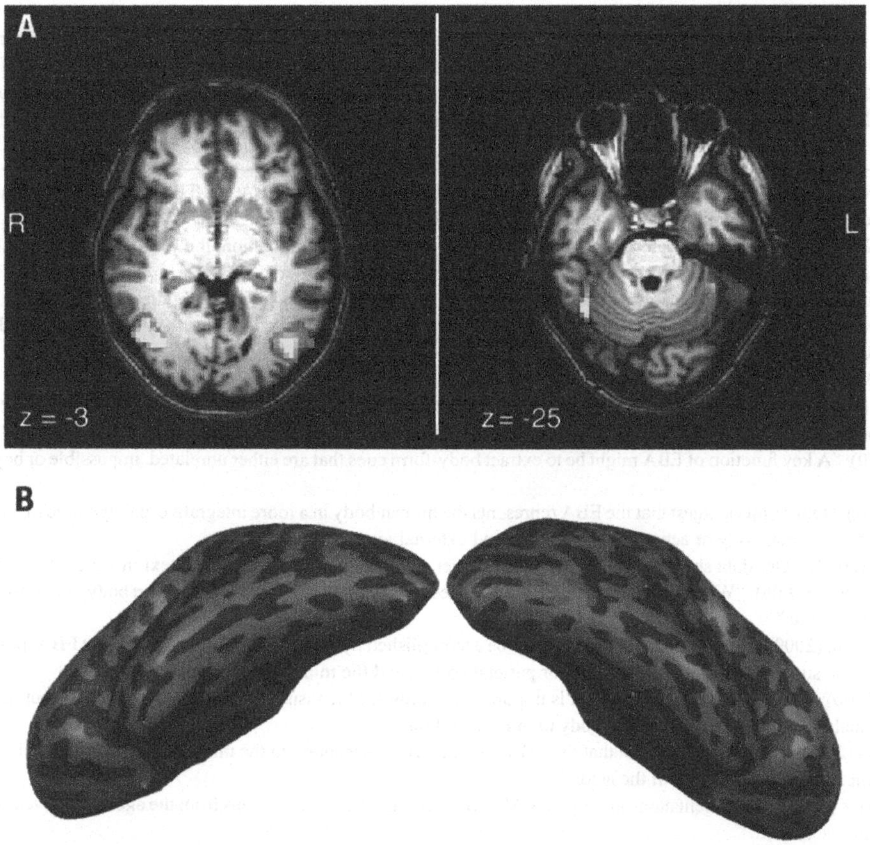

Figure 1. Location of occipitotemporal body-selective regions (extrastriate body area: EBA; fusiform body area: FBA). (A) EBA (left) and FBA (right) in one participant, defined by a contrast of bodies – (chairs + scenes), $p < .00005$. Z values refer to position of axial slices in Talairach space. (B) Results of a fixed-effects, group-average contrast of (bodies – chairs), $n = 8$, $p < .00001$, rendered on the inflated cortical surface of a single participant.

of subsequent work. This includes fMRI studies of healthy human populations and of psychiatric populations, and extends to investigations with transcranial magnetic stimulation (TMS), patients with neurological damage, event-related potential (ERP) and magnetoencephalography (MEG) techniques, and nonhuman primates.

We focus the present discussion on these two regions because they are clearly identifiable areas involved in visual body analysis, while acknowledging that they may be further subdivided by functional and/or anatomical criteria (e.g., Bracci, Ietswaart, Peelen, & Cavina-Pratesi, 2010; Weiner & Grill-Spector, 2011), and that they must operate not in isolation but rather in concert with other brain areas. (We will return to these points later.) Now that there is a rich set of data about the properties of EBA and FBA, multiple research groups have proposed that their neural activity directly underpins a plethora of functions (Table 1). These include identifying other individuals, perceiving emotions, perceiving body movements, understanding the meaning of others' actions, and even the control of motor movements.

In our view, many of these proposals extend too far past the data. Here, we critically consider current evidence on the possible function(s) of EBA and FBA, with reference to a more parsimonious model: The extrastriate and fusiform body areas jointly create a detailed but cognitively unelaborated visual representation of the appearance of the human body. This representation makes explicit the aspects of the image that contain bodies or body parts, and represents their shape and posture in some detail. It does not, we argue, make explicit high-level information about other people, such as their identities, actions, or emotional states. Thus, this representation provides a general perceptual infrastructure that contributes to the information processing in other interconnected brain areas that extract complex and often context-dependent information about the people we observe.

BACKGROUND

Downing et al. (2001) reported new evidence from human fMRI for a lateral occipitotemporal region that

TABLE 1

The discovery of highly body-selective regions in the lateral (extrastriate body area; EBA) and ventral (fusiform body area; FBA) occipitotemporal cortex has led to a wide diversity of proposals for the functional role(s) that these regions play. The quotations below are taken out of context and are simply meant to illustrate this diversity. In this article, we argue, contrary to many (but not all) of these proposals, that these regions perform a role that is largely restricted to representing the shape and posture of the body, and it is not elaborated with information about identity, emotion, motion, action goals, or motor control

- Arzy et al. (2006): "Collectively, these data show that distributed brain activity at the EBA and TPJ as well as their timing are crucial for the coding of the self as embodied and as spatially situated within the human body."
- Astafiev et al. (2004): "Our results indicate that in addition to this visual recognition function, the EBA integrates visual, spatial attention, and sensory-motor signals involved in the representation of the observer's body."
- Blanke et al. (2010): "Our data show that the EBA is also involved in mental imagery of human bodies."
- Calvo-Merino et al. (2010): "Our results suggest that the EBA and vPMC may be two complementary components of the aesthetic perception network for bodies."
- Chan et al. (2004): "We propose that the EBA plays a relatively early role in social vision."
- Costantini et al. (2005): "The higher BOLD signal during observation of impossible movements . . . not only suggests that EBA may be activated during action observation but also that this area codes body and action related stimuli multimodally."
- Cross et al. (2010): "A key function of EBA might be to extract body-form cues that are either unrelated, impossible or beyond what the viewer's body can do."
- David et al. (2007): "Our results suggest that the EBA represents the human body in a more integrative and dynamic manner, being able to detect an incongruence of internal body or action representations and external visual signals."
- De Lange et al. (2008): "Our data show that EBA activity is further influenced by the motoric context in which the body part is presented."
- Downing, Peelen et al. (2006): "We speculate that the EBA computes a static representation of the human body and is not involved in analysis of biological motion *per se*."
- Hodzic, Muckli et al. (2009): "[Identity analysis] appears to be accomplished by a network comprising the right FBA, positioned ventrally to the EBA, regions of the superior parietal lobe, the inferior parietal cortex, and the middle frontal gyrus."
- Jackson et al. (2006): "It is thus likely that the EBA is important not only for the visual processing of body parts, but also for . . . automatically mapping the visual representation of another's body to one's own body."
- Jastorff & Orban (2009): "Our results suggest that the EBA and the FBA correspond to the initial stages in visual action analysis, in which the performed action is linked to the body of the actor."
- Kable & Chatterjee (2006): "Representations in the pSTS, MT/MST, and EBA abstract actions from the agents involved and distinguish between different particular actions."
- Kokal et al. (2009): "The presence of these regions [EBA and pSTS] in our [joint action] networks suggests that the process of integrating observed and executed actions . . . may also occur at a more sensory level."
- Kontaris et al. (2009): "In contrast to pSTS, EBA and FBA are decoupled from motor systems."
- Kuhn et al. (2011): "Our study suggests that both EBA and FFA play a role in the representation of one's own body and in the control of voluntary action."
- Marsh et al. (2010): "The extrastriate body area may be geared to take into account the social meaning of actions, so that actions can be understood with reference to the person executing them."
- Moro et al. (2008): "Visual analysis of human body stimuli is based on the division of labor into two cortical systems, with EBA and FBA representing the actors' identity and vPMc mapping the observed action in a neutral format with respect to the identity of the acting bodies."
- Myers & Sowden (2008): "We argue that the right EBA may perform an important sorting of body part images by identity (including self-recognition) and may interact both with brain areas involved in sensory processing and social cognition."
- Newman-Norlund et al. (2007): "EBA ... may represent the brain basis of our ability to relate the movements of others directly with our own movements in joint-action situations."
- Pierno et al. (2009): "EBA . . . [has] a role concerned with the discrimination of specific aspects characterizing the observed actions."
- Pitcher et al. (2009): "Our results extend earlier findings by showing that rEBA represents bodies in their most common configuration."
- Pourtois et al. (2007): "[There is] a major role for EBA in the initial perceptual analysis of body shapes."
- Ramsey & Hamilton (2010): "Our data suggest that [FFA, EBA, FBA] are also recruited in more social and dynamic contexts; they distinguish between two intentional agents who are acting in a goal-directed fashion."
- Saxe et al. (2006): "[There is] a role for the right EBA in the perception of other people per se, perhaps as input to subsequent perception of others' actions and reasoning about other minds."
- Suchan et al. (2010): "Body image distortion is related at least in part to structural alteration in the EBA."
- Urgesi, Candidi et al. (2007): "Thus, the present data clearly show that EBA is crucial in processing bodily forms but not bodily actions."

responds strongly and selectively to images of human bodies and body parts. This followed earlier findings of similar face-selective (Kanwisher, McDermott, & Chun, 1997; Puce, Allison, Asgari, Gore, & McCarthy, 1996) and scene-selective (Aguirre, Zarahn, & D'Esposito, 1998; Epstein & Kanwisher, 1998) areas, and was part of a program to determine the extent of such apparently specialized regions (e.g., Downing, Chan, Peelen, Dodds, & Kanwisher, 2006). Several years after identifying EBA, we (Peelen & Downing, 2005a) identified a second highly body-selective region in the fusiform gyrus, overlapping, but distinct from, the fusiform face area (FFA). Shortly afterwards, Schwarzlose et al. (2005) reported that high-resolution fMRI revealed a double dissociation between FBA and FFA.

fMRI studies have demonstrated that EBA and FBA respond to photorealistic depictions of whole human bodies or body parts, significantly more than to faces, face parts, objects, object parts, scenes, visual motion, and other control stimuli (Downing, Chan et al., 2006; Downing et al., 2001; Peelen & Downing, 2005a; Schwarzlose et al., 2005; Spiridon, Fischl, & Kanwisher, 2006; Weiner & Grill-Spector, 2010). The responses of these regions are not limited to full-cue stimuli: Their body selectivity extends to line drawings, "stick figures," and silhouettes (Downing et al., 2001; Peelen & Downing, 2005a). The body representations in these regions appear to be viewpoint-dependent—in an fMRI adaptation design, changes in view greater than 45° result in release from adaptation to the level of a new stimulus (Taylor, Wiggett, & Downing, 2010). There is evidence from intracranial ERP for a body-selective visual response originating at the approximate site of EBA (Pourtois, Peelen, Spinelli, Seeck, & Vuilleumier, 2007), and this is confirmed by scalp ERP (Thierry et al., 2006) and MEG (Ishizu, Amemiya, Yumoto, & Kojima, 2010) studies, and near infrared spectroscopy (Ishizu, Noguchi, Ito, Ayabe, & Kojima, 2009). Notably, fMRI, TMS, and ERP studies indicate that the representation of bodies in EBA is more strongly part-based than in FBA (Costantini, Urgesi, Galati, Romani, & Aglioti, 2011; Taylor, Wiggett, & Downing, 2007; Taylor, Roberts, Downing, & Thierry, 2010; Urgesi, Calvo-Merino, Haggard, & Aglioti, 2007). Finally, recent fMRI studies of macaques show body-selective activation patches in the temporal cortex (Pinsk et al., 2005, 2009; Tsao, Freiwald, Knutsen, Mandeville, & Tootell, 2003), making an important link between human studies and previous single-unit work that reported strong selective neural responses to hands or whole bodies (Desimone, Albright, Gross, & Bruce, 1984; Gross, Bender, & Rocha-Miranda, 1969; Gross, Rocha-Miranda, & Bender, 1972; Oram & Perrett, 1996; Wachsmuth, Oram, & Perrett, 1994).

Is the response to human bodies and body parts in EBA and FBA functionally relevant for body perception? At least five published studies show a selective disruption to body- or body-part perceptual tasks following TMS over EBA (Calvo-Merino, Urgesi, Orgs, Aglioti, & Haggard, 2010; Pitcher, Charles, Devlin, Walsh, & Duchaine, 2009; Urgesi, Berlucchi, & Aglioti, 2004; Urgesi, Calvo-Merino et al., 2007; Urgesi, Candidi, Ionta, & Aglioti, 2007). These effects are specific to stimulus type and stimulation location, affecting only body perception (and not other categories) and only when TMS is applied to EBA (and not other nearby areas). Further, a recent study of neuropsychological patients (Moro et al., 2008) found that impaired performance on body perception tasks was associated with lesions to areas that were consistent with EBA and FBA, and/or connections between these two regions (see also Kemmerer & Tranel, 2008; Schwoebel & Coslett, 2005). Taken together, these findings provide evidence that the activity in EBA and FBA plays a selective and causal functional role in some aspect(s) of body perception.

EBA and FBA are found close to, or overlapping with, other known regions. Thus, previous research has had to functionally distinguish these regions from their neighbors. For example, Downing et al. (2001) reported a dissociation between the responses of EBA and nearby motion-selective region hMT+ in the responses to moving objects and bodies (see also Pitcher et al., 2009; Spiridon et al., 2006; Valyear & Culham, 2010). Similarly, Peelen, Wiggett, and Downing (2006) used multivoxel pattern analysis (MVPA) (Haynes & Rees, 2006; Kamitani & Tong, 2005; Mur, Bandettini, & Kriegeskorte, 2009; Oosterhof, Wiggett, Diedrichsen, Tipper, & Downing, 2010; Peelen & Downing, 2007b) to distinguish the responses of motion-selective hMT+ and EBA to point-light biological motion stimuli (see also Downing, Wiggett, & Peelen, 2007). Most recently, Weiner and Grill-Spector (2011) used high-resolution fMRI to anatomically dissociate hMT+ and EBA. With respect to FBA, MVPA (Peelen et al., 2006), high-resolution fMRI (Schwarzlose et al., 2005), and developmental time course (Peelen, Glaser, Vuilleumier, & Eliez, 2009) show that the body-selective and face-selective responses in the fusiform gyrus can be dissociated. Importantly, this overlap means that occipitotemporal activations in the general region of EBA or FBA can be difficult to interpret without careful localization within-studies and within-participants (Figure 2).

The overlap and close proximity of EBA and FBA to other functional regions has further implications as well. For one, the nature of the neighboring areas may provide clues about the properties of EBA and FBA. For example, the proximity of EBA to motion-selective regions suggests that it interacts with dynamic visual representations, although we argue below that the body representation in EBA itself is not dynamic. Furthermore, very recent evidence suggests that the highly body selective responses we define as EBA and FBA may be situated in a much broader, more weakly body-sensitive expanse of cortex. Specifically, Orlov, Makin, and Zohary (2010) find large occipitotemporal regions that respond subtly but reliably more to specific body parts (e.g., upper or lower limbs) than to other body parts. Both of these observations provide an important context for

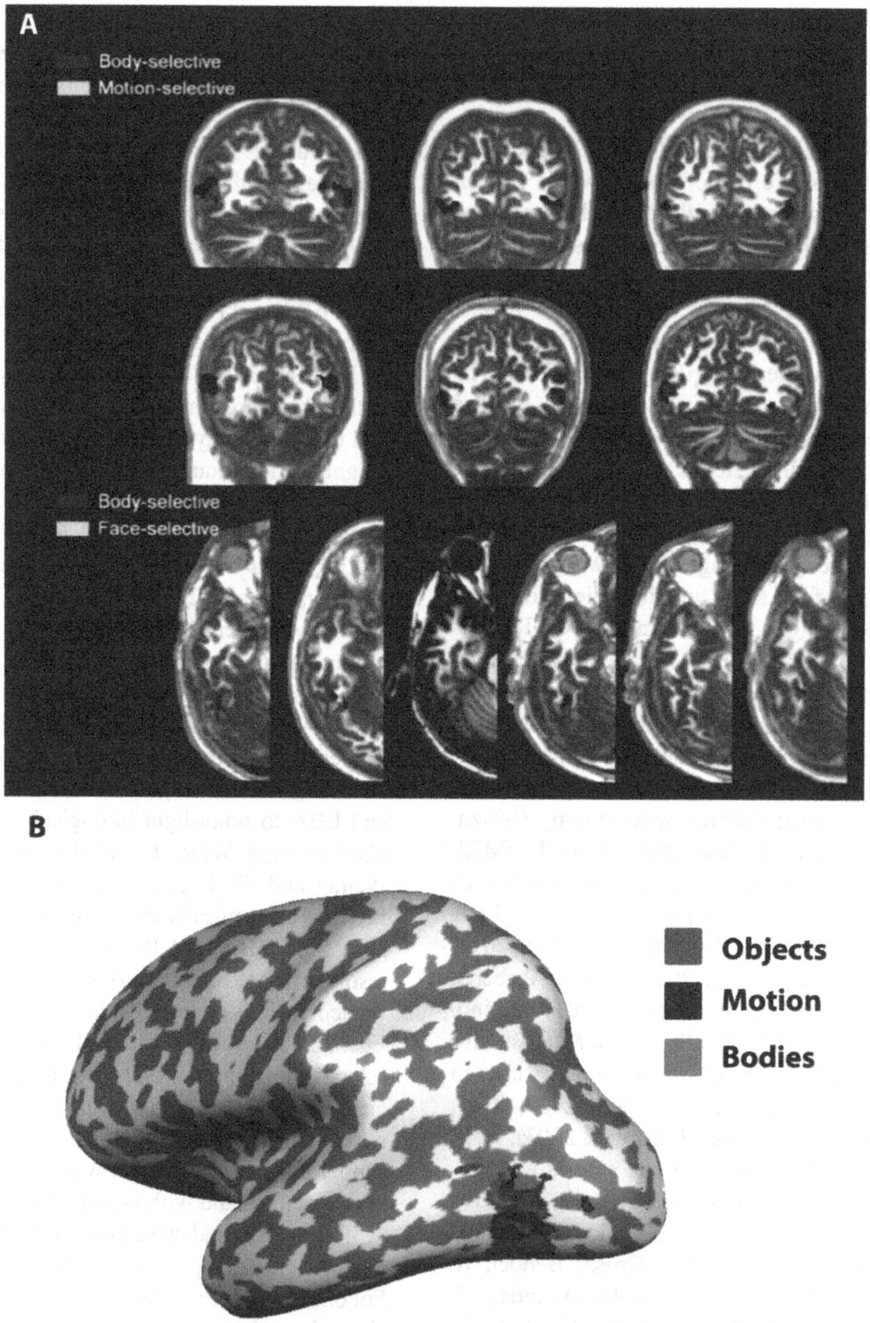

Figure 2. Overlap of functionally defined EBA and FBA and neighboring regions. (A) Overlap in six individual participants (at $p < .001$), adopted from Peelen et al. (2006). Upper panel shows overlap between extrastriate body area and motion-selective hMT+. Lower panel shows overlap between fusiform body area and fusiform face area. (B) Schematic illustration of the overlap of body-, object-, and motion-selective regions, rendered on the inflated cortical surface of a single participant. Random-effects, group-average analysis, $n = 14$, $p < .001$ (bodies), $p < .0001$ (motion), $p < .00001$ (objects). Results like these highlight the necessity for functional localization of regions of interest in studies of extrastriate visual areas.

understanding EBA and FBA and we consider them further in the concluding section.

We make a last note on interpreting fMRI results before turning to a review of the evidence. At several points in the following discussion, we attribute modulations of activity in EBA or FBA to (often unintended) variations in the degree to which selective attention is allocated to bodies or body parts in the stimulus. Many studies have shown attentional modulation in retinotopic visual cortex (Brefczynski & DeYoe, 1999; Gandhi, Heeger, & Boynton, 1999; Kastner, De Weerd, Desimone, & Ungerleider, 1998)

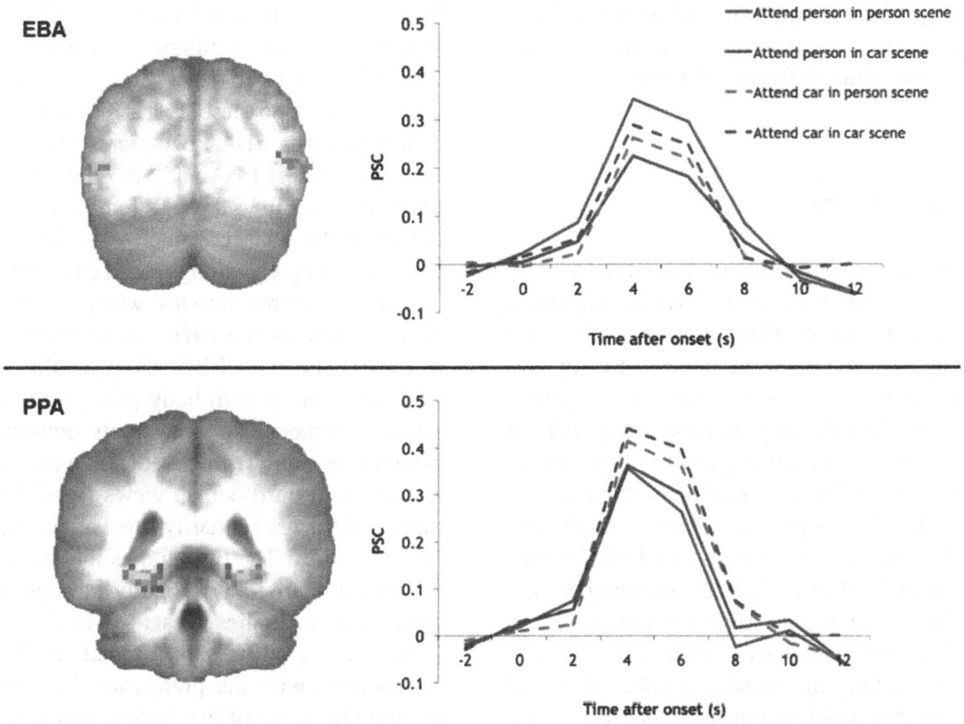

Figure 3. Event-related time courses in EBA (top) and PPA (bottom) in response to pictures of natural scenes. The scenes could contain persons or cars, and participants were instructed to attend (in different runs) to one of these categories. When attention was directed to persons (solid lines), responses in EBA showed stronger responses to scenes containing persons than to scenes containing cars (green solid line higher than red solid line). By contrast, when attention was directed to cars (dashed lines), no selective response to person scenes was observed (no difference between green dashed line and red dashed line). These results show that top-down attention can strongly modulate responses in EBA. The interaction between attention and scene content was specific to body-selective regions and was not observed in PPA (bottom). For more analyses and information, see Peelen et al. (2009).

as well as in high-level extrastriate areas such as the parahippocampal place area (PPA) (Epstein & Kanwisher, 1998), FFA, and hMT+ (e.g., O'Craven, Downing, & Kanwisher, 1999; O'Craven, Rosen, Kwong, Treisman, & Savoy, 1997; Wojciulik, Kanwisher, & Driver, 1998). On these grounds, it is highly likely that varying attention to body or body part stimuli will have congruent effects on the responses of EBA and FBA. Consistent with this, Morris, Pelphrey, and McCarthy (2006) found that the EBA activity elicited by bodies was reduced when a face was present in the image, compared to when there was not. A simple attentional account of these findings is that when a face is present, it is spatially attended (or fixated) at least some portion of the time, and the net effect on EBA response is suppressive (given the low response in this region to faces presented alone; Downing et al., 2001). Further direct evidence is found in recent data from Peelen, Fei-Fei, and Kastner (2009). In that study, participants were required to attend to either people or cars in briefly presented natural scenes. Activity in EBA, but not scene-selective PPA, was modulated by the task set, such that selectivity for bodies (vs. cars) was stronger during the body task than the car task (Figure 3). These category-based attention effects were highly robust, and were found for both spatially attended and unattended pictures. This means that where experimental manipulations confound how much spatial or nonspatial attention is focused on body stimuli, the interpretation of results may be compromised.

FIVE POSSIBLE FUNCTIONS FOR BODY-SELECTIVE EXTRASTRIATE REGIONS

The following discussion is organized into five sections, each describing a possible role of EBA and FBA in body representation: perceiving identity, perceiving emotion, perceiving body motion, perceiving actions and goals, and motor control. For many of these functions, researchers have claimed that EBA or FBA perform complex, elaborate roles (Table 1); for example, they claim that these regions "understand" others' actions, or coordinate one's own behavior with that of another individual. We will argue that the findings that were taken as support for these claims can be more parsimoniously explained by a model in which EBA and FBA represent visual features of the body such as shape and posture without further elaboration.

Perceiving identity

Although person identity is often perceived from facial cues, the rest of the body also provides important information about identity. Body posture, shape, and gait can all be used to determine person identity, and may be particularly useful when the face is poorly visible such as when viewing someone from behind or at a distance. Since recognizing identity from bodies requires the perceptual analysis of body stimuli, it is likely that EBA and FBA play an important role in this process. Below, we discuss studies that have investigated the role of EBA and FBA in extracting person identity, and indicate how a perceptual account of these regions can accommodate these findings.

Several studies have investigated whether EBA and FBA respond differentially to images of one's own versus another person's body, as this is perhaps the most fundamental identity distinction. Chan, Peelen, and Downing (2004) reported no difference between EBA responses elicited by viewing the bodies of the self versus familiar others in a blocked design fMRI experiment. Subsequent studies that contrasted the self with familiar others similarly found no difference in EBA or FBA (Devue et al., 2007; Hodzic, Kaas, Muckli, Stirn, & Singer, 2009), although one study (that did not localize EBA/FBA) reported a broad swath of occipitotemporal cortex that responded more to self than to familiar other images (Sugiura et al., 2006). The results of studies that contrasted images of the self with *unfamiliar* others are less consistent: Hodzic, Kaas et al. (2009) found no difference in right EBA, but a stronger response to self than other bodies in left EBA and right FBA. However, in another study, these authors did not find a difference between self and unfamiliar other in EBA or FBA (Hodzic, Muckli, Singer, & Stirn, 2009). Finally, Vocks et al. (2010) reported small but significant increases to own bodies relative to unknown others' bodies in right EBA and FBA.

Thus, while most studies reported no difference in EBA and/or FBA between self and other bodies, some studies reported modest increases in one or both of these regions. It should be noted that these studies compared the overall BOLD signal in EBA and/or FBA to images of the self versus a familiar or unfamiliar other. It is unlikely, however, that a region that extracts identity from perceived bodies would show gross differences in the overall response magnitude to different identities. More likely, different identities would elicit different patterns of activity resulting in roughly the same gross level of response (Kriegeskorte, Formisano, Sorger, & Goebel, 2007). So these studies may not be ideally suited to test whether EBA/FBA represents identity. Rather, they test whether EBA and FBA are primarily involved in representing the self (e.g., for guiding actions) or the other (e.g., for social cognition)— and their results do not point to a uniquely strong role in either process.

Another approach to test whether EBA and/or FBA primarily encode the self or the other is to test the effect of viewpoint: If EBA/FBA would preferentially respond to one's own body parts, we would expect a stronger response to body parts presented in a first person viewpoint (camera at eye level), while stronger responses to third person views would be expected if these regions are primarily involved in the encoding of others' bodies. The two studies that have manipulated viewpoint both found slightly stronger responses in right EBA to bodies presented in third person view (Chan et al., 2004; Saxe, Jamal, & Powell, 2006), inconsistent with the preference for one's own body reported by some of the studies reviewed above.

Using a blocked-design adaptation approach, Myers and Sowden (2008) reported significantly stronger responses (i.e., less adaptation) in right, but not left, EBA (FBA was not tested) to blocks in which images of the participants' own hand were alternated with images of an unknown other's hand, relative to blocks where two other individuals' hands were alternated. The authors concluded that right EBA contains different neural populations to represent the appearance of one's own versus others' body parts. These results may alternatively be explained by differences in attention: It is not unlikely that participants paid more attention to their own hands than to a stranger's hand, especially because the participants' own body parts appeared relatively infrequently in the experiment. Such attention effects would be expected to primarily modulate higher-level visual areas strongly responsive to the hand stimuli, rather than weakly responsive areas or lower-level visual cortex.

Perceiving identity is, of course, not limited to distinguishing the self from others. Surprisingly, only a few studies have investigated the role of EBA or FBA in representing body identity beyond the self/other distinction. Kable and Chatterjee (2006) used a long-term, event-related adaptation design, with dynamic whole-body/face stimuli, to investigate adaptation to actor identity. No significant identity adaptation was found in EBA. In a subsequent extension of this study, the absence of significant identity-related adaptation effects in EBA and FBA was confirmed (Wiggett & Downing, 2010). Ramsey and Hamilton (2010) also used an adaptation design with the similar aim of examining representations of actors. Functional localization was not performed, but anatomical regions of interest around fusiform and middle temporal gyri (and other areas) were tested with small-volume correction analyses. Both areas showed significant adaptation for repeated versus nonrepeated actors, which the authors attributed to "person identity processes" taking place in EBA and FBA. However, it is not clear which aspect of the stimulus drove this adaptation—whether it was the repetition of actor identity, the repetition of body shape, or the repetition of low-level visual aspects such as the color of the clothes.

Together, these results suggest that EBA and FBA are involved in the encoding of body parts belonging to the self as well as to others. Small modulations reported in some of the studies reviewed here may be explained by attentional differences—looking at a picture of yourself may simply be more interesting than looking at a picture of a stranger. Notably, identity effects tend to be found when identity is either an explicit part of the participants' task (Hodzic, Kaas et al., 2009; Ramsey & Hamilton, 2010; Sugiura et al., 2006) or else is made explicit to participants by direct cues (Vocks et al., 2010). This pattern points strongly to the idea that identity modulation in EBA or FBA is the result of top-down influences, including attentional biases.

What could be the role of EBA and FBA in the perception of body identity? We argue that the role of these regions consists in creating a perceptual representation of the shape and posture of the body and its parts, which may then be used by other brain regions (e.g., the anterior temporal lobes; Kriegeskorte et al., 2007) to represent person identity explicitly. On this account, EBA and FBA encode body identity only in the sense that these regions differentiate between individuals with different body shapes. We argue that EBA and FBA do not, however, explicitly represent person identity beyond body shape, and as such may not specifically differentiate between the bodies of the self and others. While identity can be extracted from body shape, identity and shape can be dissociated (people age, grow, lose or gain weight, undergo plastic surgery, etc.). Furthermore, identity can be extracted from multiple body parts (e.g., lower and upper body). We expect responses in EBA/FBA to follow the perceived body shape (or body part) rather than the perceived body identity.

Note that our account does not exclude the possibility that responses in EBA and FBA can be influenced by identity through top-down modulation: The sight of a romantic partner (or, indeed, the self) may increase responses in EBA/FBA due to increases in attention and arousal. Also, we expect that familiarity with particular body shapes (e.g., growing up in a country with tall people) may change the tuning of body-selective neurons, such that their discrimination of body shapes is optimal for the current environment. Finally, the knowledge of another person's body shape (available after identifying the person, whether through visual or nonvisual cues) may modulate responses in EBA and FBA, as, for example, in cases where clothes obscure most of the body. In other words, responses in EBA and FBA are not merely a copy of the visual image, but more closely correspond to the subjective percept, which is influenced by identity (and other factors). The critical test of our account would require an experiment that dissociates body shape and body identity in combination with suitable techniques (e.g., MVPA). Distributed activity in EBA and FBA should conform to a "space" in which different body shapes/parts, but not necessarily different individuals, are represented by systematically different corresponding patterns.

Perceiving emotion

Basic emotions, such as fear, anger, and happiness, are associated with characteristic body postures and movements (Atkinson, Dittrich, Gemmell, & Young, 2004). Apart from providing information about others' emotional states, bodies may also evoke emotions in the observer—for example, through empathy (e.g., with others' pain), admiration of beauty (e.g., in dance), or sexual arousal. What is the role of occipitotemporal body-selective areas in these processes? Here we consider the evidence from studies that have implicated EBA or FBA in the processing of emotion from body stimuli.

Research on emotional body perception has largely addressed questions that had earlier been addressed in the field of emotional face perception, such as comparing fMRI responses to emotional expressions with responses to neutral expressions. Employing this approach, research on face perception indicates that emotional expressions modulate responses in visual cortex, particularly the fusiform face area (for a review, see Vuilleumier, 2005). These emotional modulations are similar to attentional modulations but are thought to be mediated by the amygdala, rather than by frontoparietal attention networks (Amaral, Behniea, & Kelly, 2003; Morris et al., 1998; Vuilleumier, Richardson, Armony, Driver, & Dolan, 2004). Such activity increases in visual cortex may correspond to the enhanced perceptual processing of emotionally salient stimuli (e.g., Anderson, 2005).

Similar to emotional faces, emotional bodies have been shown to modulate perceptual responses, with increased visual cortex (and amygdala) activation to arousing body expressions such as fear (Hadjikhani & de Gelder, 2003). A large number of studies from multiple laboratories have replicated these effects, using various emotions and both static and dynamic body stimuli (e.g., Carter & Pelphrey, 2008; Flaisch, Schupp, Renner, & Junghofer, 2009; Grezes, Pichon, & de Gelder, 2007; Grosbras & Paus, 2006; Kret, Pichon, Grezes, & de Gelder, 2011; Nummenmaa, Hirvonen, Parkkola, & Hietanen, 2008; Peelen, Atkinson, Andersson, & Vuilleumier, 2007). Although it is beyond the scope of this paper to review these studies in depth, a few general observations emerge from this literature. For example, it seems that arousal rather than valence is driving emotional modulation, since no modulation was found for bodies expressing sadness (Peelen et al., 2007). Further, emotional modulation has been primarily, but not exclusively, reported in body-selective parts of visual cortex (i.e., at the location of EBA and FBA), with some studies (mostly those using static body postures with blurred faces) finding stronger modulation in fusiform gyrus (presumably FBA) than at the location of EBA (e.g., Van de Riet, Grezes, & de Gelder, 2009). In a study using dynamic body expressions, it was shown that emotional modulation is proportional to static body selectivity (measured in a separate localizer scan) across the voxels of EBA and FBA, with the strongest modulation observed in voxels with the strongest body selectivity (Peelen et al., 2007). In the same study, emotional modulation in EBA and FBA was shown to correlate with emotion-driven increases in the amygdala, again highlighting the similarity between the mechanisms involved in emotional body and face processing (Peelen et al., 2007). Finally, modulation in visual cortex (including putative body-selective regions) has also been reported for sexually arousing body stimuli (Cikara, Eberhardt, & Fiske, 2011; Hamann, Herman, Nolan, & Wallen, 2004; Mouras et al., 2003; Ponseti et al., 2006).

Together, these results suggest that activity increases in EBA and FBA in response to emotional or sexual body stimuli primarily reflect modulatory effects (or "emotional attention"), rather than a direct encoding of emotions within these regions. It should be noted, however, that (as far as we know) no study has directly tested for explicit representations of emotions in these regions, as by using fMRI adaptation or MVPA (cf. Peelen, Atkinson, & Vuilleumier, 2010).

Observing someone else's body in acute physical pain can hurt, especially when this person is a loved one. This empathy for others' pain has been shown to activate regions in the anterior insula and anterior cingulate cortex, regions that also activate when one is directly experiencing pain (Morrison, Lloyd, di Pellegrino, & Roberts, 2004; Singer et al., 2004; but see Morrison & Downing, 2007). Although body-selective regions have not been implicated in the direct experience of pain, they may play a role in the perception of pain in others. For example, Singer et al. (2004) found activity in fusiform gyrus as well as more posterior visual regions in response to observing another's hand that is thought to be in pain (as indicated by a symbolic cue) relative to when this hand was known not to be in pain. Similarly, Gu et al. (2010) found increased responses in occipitotemporal regions (probably including EBA) to photographs of body parts in painful daily-life incidents. However, such pain-related modulation was not observed in several other studies that showed a needle or knife touching a hand or foot (Benuzzi, Lui, Duzzi, Nichelli, & Porro, 2008; Morrison et al., 2004; Morrison, Peelen, & Downing, 2007). The only study to directly investigate pain-related responses in functionally localized EBA also used images showing a needle contacting a hand. No modulation was found in EBA, and it was concluded that the EBA is not relevant for empathy (Lamm & Decety, 2008). Thus, although there is currently no direct evidence that body-selective regions are modulated by observed pain, several studies suggest this might be the case for some stimuli. Studies that used stimuli in which a noxious object interacts with a body part (e.g., a needle contacting a hand) typically failed to find modulation in visual cortex. A possible explanation could be that in these studies participants directed attention toward the noxious object, or toward the local site of impact, rather than the body part. Such differences in attentional focus could have offset any pain-related modulatory effects.

There is a strong emotional aspect to clinical eating disorders such as anorexia nervosa and bulimia nervosa, in that patients may react with strong negative feelings toward the appearance of their own bodies or to others they perceive as overweight. These disorders can be associated with distortions of the body image, as revealed in misperceptions of body shape and size, and hence some authors have made a connection between these syndromes and EBA or FBA. Uher et al. (2005; see also Aleong & Paus, 2010; Sachdev, Mondraty, Wen, & Gulliford, 2008) tested visual responses to body images in patients and in control participants. Line drawings of women in either underweight, normal, or overweight versions were compared in a blocked-design fMRI experiment. Low body selectivity was seen in occipitotemporal cortex in patients relative to controls. Furthermore, in a fusiform region, patients showed reduced activity (relative to controls) for the overweight stimuli, and increased activity for the underweight stimuli. In a study of functional and structural neuroanatomy, Suchan et al. (2010) identified EBA in anorexia patients and healthy controls, who also performed a behavioral test in which they estimated their own body sizes against a set of standardized silhouettes. Whole-brain voxel-based morphometry showed a reduction in gray matter volume in the patients, relative to controls, narrowly restricted to a region that fell in the functionally defined left EBA. Further, there was a negative relationship between anatomy and behavioral performance: Greater error on the body size judgment task was related to lower gray matter volume in EBA. Together, findings such as these suggest that eating disorders may be associated with distorted perception of bodies, or reduced or altered attention to bodies, and these characteristics may be related to differences in structural neuroanatomy in or around EBA.

Finally, a recent study used TMS to investigate the role of EBA in the aesthetic evaluation of dance postures (Calvo-Merino et al., 2010). TMS over EBA reduced aesthetic sensitivity (i.e., aesthetic judgments were less consistent) for body stimuli, but not for control stimuli, relative to TMS over ventral premotor cortex. Since the differences between the dance postures were related to changes in limb position, a reduction of aesthetic sensitivity would be expected if the perceptual processing of limb position was disrupted by TMS over EBA. As such, this study does not provide evidence that EBA is specifically involved in aesthetic evaluation beyond the perceptual representation of body posture.

To conclude, there is convincing evidence that emotional and sexual body stimuli (vs. neutral bodies) can modulate processing in body-selective regions in occipitotemporal cortex. These modulatory effects of emotion are likely to be similar to the effects of attention, in that they increase the strength of perceptual representations, but may have a different cortical origin than visuospatial attention effects. No evidence currently exists for a role of EBA or FBA in the direct representation of emotions beyond the perceptual representation of body shape and posture.

Perceiving body motion

The human body is nearly always moving, and this results in characteristic patterns of visual motion. So it is to be expected that visual motion will be a critical component of the neural representation of the body. Indeed, given the close proximity of EBA to motion-selective area hMT+ (Downing et al., 2007), a plausible account of body-selective responses in EBA would be that these reflect selectivity to body movements (either implied by the stimulus or directly depicted) rather than body shape. Perhaps counterintuitively, then, there is accumulating evidence that the body representations in EBA and FBA are not dynamic: that is, they do not explicitly represent the motions of body parts.

Downing et al. (2001; see also Beauchamp, Lee, Haxby, & Martin, 2002) manipulated the combination of object kind and motion cues in order to test whether EBA shows a particularly strong response to characteristic human motions. Objects and bodies were presented either statically, with simple rigid motion, or with realistic dynamic motion. In EBA, increasing responses were seen from static to rigid to dynamic body stimuli, but the same pattern was found for objects as well, suggesting a general effect of motion complexity that was not specific to body processing.

Rather than use full-cue stimuli, however, much of the work on perception of human motion has made use of variants on Johansson's (1973) "point-light" stimuli, which can convey the patterns of biological motion in the absence of surface form (see Blake & Shiffrar, 2007). Grossman and Blake (2002) reported that EBA did not respond selectively to point-light displays of various whole-body movements, relative to scrambled controls. However, in later experiments, such effects were demonstrated and were extended to FBA (e.g., Michels, Kleiser, de Lussanet, Seitz, & Lappe, 2009; Michels, Lappe, & Vaina, 2005; Peelen et al., 2006; Peuskens, Vanrie, Verfaillie, & Orban, 2005). For example, Peelen et al. (2006) found greater responses in the fusiform gyrus and inferior temporal sulcus to intact versus scrambled whole-body point-light actions. Using MVPA, we showed that the degree of selectivity for these animations correlated positively, voxel-by-voxel, with the degree of selectivity for static body images in these regions (and not with motion or face selectivity). We argued that the "biological motion" response of EBA and FBA should be attributed to the percept of a body *per se* (which is elicited by structure from motion in intact point-light animations), and not to an explicit representation of patterns of biological motion.

Jastorff and Orban (2009), however, have argued to the contrary. They tested the responses of EBA and FBA to point-light animations of whole-body movement in a 2 × 2 design. The individual points moved either with realistic patterns of biological movement, or translated rigidly as a group. Further, they were either configured to conform to the shape of a human body (e.g., points at key joints) or not. In EBA and FBA, an interaction was found: responses were maximal when the motion type was biological and the configuration of points was intact. Jastorff and Orban (2009) argued that form and motion interact in these regions; that is, that the neural representations in EBA and FBA encode dynamic aspects of movement explicitly. However, it is only in the intact configuration/intact motion condition that body structure is apparent. Without realistic patterns of biological movement, the body-like structure implicit in a rigidly moving pattern of points is very weak. On these grounds, the results of Jastorff and Orban (2009) are instead consistent with our proposal that it is body structure, rather than biological motion *per se*, that is encoded in EBA and FBA.

Static stimuli can strongly imply motion and can engage motion-processing mechanisms: Consider a snapshot of a sprinter taken as she bursts from the starting blocks. This effect is captured in the "representational momentum" phenomenon (Kelly & Freyd, 1987), in which static images of an object in motion are subsequently recalled as if the object had continued traveling in the depicted direction. In a comparison of "active" (e.g., alert and ready to perform a movement) and "passive" (e.g., resting) whole-body postures, Downing et al. (2001) found no modulation of activity in EBA. The same manipulation, however, significantly modulates activity in nearby hMT+ (Kourtzi & Kanwisher, 2000; but see Lorteije et al., 2010) suggesting that that region, but not EBA, is sensitive to implied motion.

In sum, we argue that the body representations in EBA and FBA are related to the shape of the body and not the visual stimulation produced by its movements. Shape information can be extracted from biological motion cues, thereby activating EBA and FBA when the perceived shape is a body or body part. Explicit representations of human body movements are extracted elsewhere in the brain. Extensive evidence from many methods points to the superior temporal sulcus as a key region for representing the patterns of human movement and perhaps for extracting meaningful cues such as direction of attention or intentions (Allison, Puce, & McCarthy, 2000; Puce & Perrett, 2003; Saxe, Xiao, Kovacs, Perrett, & Kanwisher, 2004).

Given that the body is nearly always in motion, and that it moves in such regular, characteristic, and often meaningful ways, it remains puzzling that static representations of the body form should be neuroanatomically distinct from the regions that extract dynamic information. Notably, distinct static and dynamic representations (of body shape and characteristic movements, respectively) have been incorporated into formal models of biological motion perception and gesture understanding (Giese & Poggio, 2003; Lange & Lappe, 2006; Peigneux et al., 2004). A common element of these models is that known body movements may be represented by static "snapshots," representing the elemental postures that characterize a movement (see also Singer & Sheinberg, 2010, for evidence from macaques). Interestingly, some computer animation software creates realistic human movement by allowing the user to specify static key frames and then rendering the movements by interpolating between these. There may be a useful analogy in the coarse-coding principle (Hinton et al., 1986), under which, for example, a wide range of line orientations can be coded efficiently by a weighted combination of detectors optimized for only a small set of orientations (e.g. horizontal and vertical). Similarly, the range of human movements may be usefully coded by reference to a "library" of commonly seen static poses that can be used combinatorially.

New experiments could test whether some postures are more likely (on behavioral or computational grounds) to act as "key frames" in the perception of movement, and whether these poses are represented specially in EBA and FBA. Furthermore, the results from studies of static images (Taylor et al., 2007) suggest a bias for local body parts in EBA relative to larger assemblies of parts in FBA, leading to the prediction that these areas contribute differently to representing movements of parts and wholes. Finally, computational studies will need to explicitly compare families of models that do or do not include static body representations, in order to see how they fare at predicting behavioral performance, and the activity in EBA and FBA, on relevant tasks.

Perceiving actions and goals

Perception of the dynamic aspects of human motion can be an end in its own right, or an initial step in the more general process of understanding the meaning, intentions, and goals behind others' actions (e.g., Grafton & Hamilton, 2007). However, understanding the actions of others does not necessarily require dynamic motion cues, as actions can be rapidly understood from static images as well (Bach, Peelen, & Tipper, 2010; Johnson-Frey et al., 2003). Thus, although occipitotemporal body-selective areas do not explicitly encode body motion, they might nonetheless contain representations of others' actions. To what extent are these areas involved in differentiating body actions? And do they take into account the surrounding context (e.g., objects, other people) of an action, or the inferred goals of the actor, when creating a body representation?

Urgesi, Candidi et al. (2007) presented static body part stimuli in a delayed match-to-sample task. The images could vary either in the action they implied (subtle variations in the position of appendages) or form (subtle variations in their shape). TMS over EBA impaired performance on the form task, but not the action task, while TMS over ventral premotor cortex produced the opposite pattern (see also Pitcher et al., 2009). These findings suggest that EBA does not critically contribute to representing actions implied from static images. This is consistent with the findings of Downing, Peelen, Wiggett, & Tew (2006), who broke movies of whole-body movements into static frames and manipulated the order in which these were presented. In the coherent condition, the frames were presented in the original order, to create a meaningful action sequence without motion cues. In the incoherent condition, frames from different actions were mixed, leading to a disjointed percept and to relatively large frame-to-frame differences. Posterior STS and fronto-parietal areas responded more to the coherent series, suggesting they represented the unfolding coherent actions. However, EBA (and to some extent FBA) responded more strongly to the incoherent than to the coherent series. These results again suggest that EBA and FBA do not form an integrated representation of body actions.

Many everyday actions are object directed, as in the use of tools or other implements. Does activity in EBA or FBA reflect the relationship between the movement and the object? De Lange, Spronk, Willems, Toni, and Bekkering (2008) found increased bilateral activity in a region they attributed to EBA (not localized) when participants observed atypical versus typical object-directed actions (e.g., drinking from a cup with a typical grasp or a clumsy, backwards grip). They argue that this shows EBA is sensitive to the "motoric context" in which body parts are viewed. When Valyear and Culham (2010) took a similar approach—comparing the observation of typical versus atypical grasps of tools—they found a left posterior middle temporal gyrus region that was sensitive to grasp. Notably, this effect did not extend to functionally defined EBA. In that study, however, both types of grasp did produce a greater response in right EBA than a control condition that showed an object simply being touched. Likewise, Pierno et al. (2009) found an increased occipitotemporal response to grasping (over pointing) movements, which they attributed to EBA. Both effects could simply reflect competition between stimuli—in grasp, but not point, conditions, less of the target object is visible, and hence that object makes a weaker competitor for attention with the hand stimulus. Moreover, grasp actions may direct attention to the object more strongly than pointing (Bekkering & Neggers, 2002), with the same result on activity in EBA.

Several recent studies compared the effects of attending to different aspects of actions. For example, Spunt, Satpute, and Lieberman (2011) had participants view movies of short, object-directed actions under different task instructions. Increased activity was seen in left lateral occipitotemporal cortex when participants covertly attended to the mechanics of the actions ("how") rather than to the descriptive ("what") or intentional ("why") aspects. The authors argue (and we agree) that this could reflect increased attention to the body under "how" instructions (since perceiving the body accurately is essential to understanding the mechanism of an action). In contrast, understanding "why" an action takes place will naturally require wider attention to other aspects of the surrounding scene.

A visual-attention account does not explain, however, why similar lateral occipitotemporal activations are found in studies that use purely verbal material. Such studies aim to test an "embodied" view of action knowledge, whereby even abstractly specified concepts (e.g., words about actions) are rooted in the activity of visuomotor brain areas. Spunt, Falk, and Lieberman (2010) had participants make "how" (relative to "why") judgments about actions described verbally. Likewise, Van Dam, Rueschemeyer, and Bekkering (2010) compared the reading of verbs that were generic with respect to a motor program (e.g., "to clean") and verbs that were more closely tied to actions (e.g., "to wipe"). And Rueschemeyer, Pfeiffer, and Bekkering (2010) compared nouns related to actions made toward the body (e.g., cup, comb, sweater) with "world-" related items that typically involve actions away from the body (e.g., wrench, pen, frisbee). All of these studies identified lateral occipitotemporal activations (variably on the left or right), and these were attributed directly to EBA by Van Dam et al. (2010) and Rueschemeyer et al. (2010).

Such results could reflect semantic or crossmodal action or body knowledge in EBA. Alternatively, they could reflect top-down activation of perceptual body representations—for example, through mental imagery or automatic association processes—after the meaning of the words has been processed elsewhere. Furthermore, the relationship of these activations to EBA is unclear, as none of these studies localized this area functionally—and the candidate regions in Van Dam et al. (2010; Figure 1, Table 3) make a particularly poor match to EBA on anatomical grounds. This is important, as noted above and illustrated in this context by the findings of Valyear and Culham (2010), because without precise localization, EBA may be confused with nearby areas with entirely different functional properties (see also Takahashi et al., 2008). Results of a recent study examining the lateral occipitotemporal responses to reading various categories of words (including nouns and verbs) emphasize this point. Bedny, Caramazza, Grossman, Pascual-Leone, and Saxe (2008) found that these verbal stimuli deactivated EBA (which was independently localized) relative to baseline, and further that there was no differentiation among word categories in that region.

All of the above studies measured gross changes in the level of activation in occipitotemporal cortex. The logic of repetition suppression (e.g., Grill-Spector, Henson, & Martin, 2006) may be more suited to detecting neural populations that encode perceived actions. This approach was first applied to EBA by Kable and Chatterjee (2006), who found that activity in a number of areas, including functionally defined EBA, was reduced when participants viewed a short movie of an action that had been seen previously in the experiment, compared to a new action. This repetition suppression effect was seen even when the actor involved in the two presentations was different, suggesting adaptation of an action representation at a relatively generic or semantic level. However, in a replication and extension, Wiggett and Downing (2010) found that this action priming effect was widespread—extending across face, object, scene, and body selective regions of interest in the lateral and ventral occipitotemporal cortex—and could not be attributed specifically to body-specific regions. Instead, it may reflect more general consequences of repetition, such as reduced attention arising from familiarity.

In summary, as we consider increasingly abstract aspects of actions, such as the objects they are directed to, their meaning, and the purpose of the actor in performing them, the relevant neural codes naturally move further away from a specific description of the posture and shape of the actor. Very similar movements can carry out very different goals (e.g., a gentle touch vs. a nasty poke), and, conversely, very different movements can achieve the same outcome (e.g., Umilta et al., 2008). When stimulus and attentional differences, and functional localization, are taken into account, we find little evidence to support the proposal that the representations of EBA and FBA can make the kinds of distinctions necessary to support high-level action understanding. Rather, as many authors have described elsewhere (e.g., Van Overwalle & Baetens, 2009), the evidence points to prefrontal and especially parietal regions as key for this process. It is likely that the information provided by EBA and FBA—particularly about body posture (Downing, Peelen, et al., 2006)—provides an important part of the input to these areas, which can in turn integrate this with input from other neural sources in order to incorporate the contextual information that is key to understanding actions.

Motor control

Movements of the body are coordinated by the joint activity of the frontal and parietal cortices and the cerebellum. However, the finding that extrastriate occipitotemporal areas respond to images of the body naturally raises the question of whether these areas also play some role in motor control (Downing et al., 2001). While they are unlikely to directly control the activity of muscles, these regions could be critically involved in visuomotor coordination by providing online visual tracking of the positions of one's own limbs. Alternatively, signals from the motor system could directly modulate the responses of neurons in these regions, perhaps to anticipate the visual consequences of body-part movements. Thus, the strongly visual response of EBA and FBA does not rule out *a priori* a role in motor behavior, or interactions with the motor system.

The first direct test of a motor role for EBA was conducted by Astafiev, Stanley, Shulman, and Corbetta (2004). Unseen, visually cued movements of the hands and feet (but not shifts of attention or the eyes) were found to increase EBA activity. This finding has been taken by others to show that EBA has a role in motor control and even to show that it is part of the human "mirror system." In a subsequent replication, we (Peelen & Downing, 2005b) questioned the involvement of body-selective neurons in unseen hand movements: In our study, EBA only partially overlapped with the region activated by unseen hand movements, and MVPA in the overlap region showed no relationship between body and movement selectivity. In a response, Astafiev, Stanley, Shulman, and Corbetta (2005) showed that the overlap between EBA and the movement-related region depended on the control condition used: The overlap was substantial when contrasting hand and foot movements with a fixation baseline, but small when contrasting these movements with a matched control condition (covert detection without limb movements). Thus, the selective activation to body-part movements appears to be localized adjacent to EBA, at least when contrasted with a matched control condition. Nonetheless, the close proximity of these regions suggests a possible interaction between motor control and visual body processing.

Several authors have proposed that EBA has a motor role, in that its activity may be modulated by the output of the motor system. Specifically, it may receive an "efference copy" of motor commands in order to anticipate the perceptual consequences of movement. For example, David et al. (2007; see also Yomogida et al., 2010) adopted such an account for an fMRI study in which participants controlled a simple visual cursor with unseen hand movements. On some trials, the visual feedback was manipulated so that it did not correspond to the participants' movements. EBA activity was greater on such trials, compared to when the feedback was correct. David et al. (2007) argued that EBA not only represents visual stimulation but also integrates visual and motor signals. Furthermore, David et al. (2009) used TMS combined with a similar protocol to show that stimulation of EBA increased response times on the detection of incorrect-feedback trials. Closer examination (described in detail in Kontaris, Wiggett, & Downing, 2009) suggests, however, that TMS stimulation may have also influenced performance on a motion task that was used as a control, and hence these results may have reflected disruption of adjacent hMT+. This possibility is relevant, given that the visual stimuli in these tasks were not body parts but simple visual cursors, and that hMT+ has itself been implicated in motor-visual mismatches (e.g., Whitney et al., 2007).

Kokal, Gazzola, and Keysers (2009) proposed that EBA is involved in integrating seen and performed joint actions in a different way—by coordinating one's own behavior with that of another individual (a review by Newman-Norlund, Noordzij, Meulenbroek, & Bekkering, 2007, makes a very similar proposal). In the key conditions of Kokal et al. (2009), the experimenter and the participant used finger movements in order to jointly create a specific angle on a clock-like device. This task produced more activity in lateral occipitotemporal cortex (EBA was not localized) than conditions in which either the participant or the experimenter alone moved the shared object. The authors argued from this finding that EBA has a role in "integrating observed and executed actions." A simpler alternative account, however, is that observing two finger movements produces more activity in EBA (or nearby hMT+) than one finger movement.

To test the proposal that EBA and FBA receive feedback from the motor system about ongoing movements, Kontaris et al. (2009) showed participants movies of moving hands under varying conditions. In the "compatible" condition, the moving hand was a live on-screen view (via MR-compatible camera) of the participants' own movements. In the "incompatible" condition, the same stimuli were shown (having been recorded in previous blocks) while the participant made similar but incompatible hand movements. Regions receiving motor information about ongoing hand movements would be expected to respond differently in these conditions—for example, by showing a reduced response to the expected, self-generated visual stimuli seen in the "compatible" condition. Consistent with this, activity in posterior superior temporal sulcus (pSTS) was suppressed to below-baseline levels by the participants' own movements in that condition (cf. Leube et al., 2003). In contrast, EBA and FBA responded strongly, but equally, to the compatible and incompatible conditions, and to a vision-only condition during which the same movies were shown without concurrent motor actions. This finding is not consistent with the idea that EBA or FBA receive an "efference copy" of motor commands or are otherwise involved in motor control.

Numerous studies have examined the brain areas involved in motor imitation, some of which have implicated EBA. For example, Jackson, Meltzoff, and Decety (2006) found extensive occipitotemporal activity for imitated versus passively observed movements, attributing this to EBA. Based on these findings (and the study by Astafiev et al., 2004), they argued that EBA automatically maps the actor's body parts on to those of an observed person, in order to support creation of an action plan. However, it seems very likely that imitation requires closer attention to the body stimuli than passive viewing, and that this explains the modulation. Indeed, Chaminade, Meltzoff, and Decety (2005) made just such an interpretation of a similar finding of modulation by imitation in occipitotemporal cortex. More recently, Molenberghs, Brander, Mattingley, and Cunnington (2010) have examined imitation of hand movements with stricter criteria, by identifying areas in which imitation produced greater activity than both passive viewing and action execution conditions. They found such activity only in pSTS.

Finally, motor imagery may produce activity in EBA and/or FBA. For example, Kuhn, Keizer, Rombouts, and Hommel (2011; see also Piefke et al., 2009) found that preparing to make either a limb movement or a face movement modulated EBA and FFA in the expected directions, even though these body parts were not in view. A difficulty of findings of this sort, however, is that they are consistent with multiple views of EBA/FBA functions: They could reflect involvement of these regions in the motor aspects of motor imagery, or simply visual imagery for the relevant body parts (see Blanke, Ionta, Fornari, Mohr, & Maeder, 2010). It is essentially the latter interpretation that Kuhn et al. (2011) adopt for their findings.

We find that the evidence does not support a direct role for EBA and FBA in the motor system. However, there is evidence in the studies reviewed above, and others, that occipitotemporal regions play some role in aspects of motor behavior. Activations are found in this region in response to both intransitive and transitive movements. Orlov et al. (2010) found small but systematic biases across a wide swath of occipitotemporal cortex for visual depictions of different body parts, including but extending well beyond EBA and FBA. Most notably for present purposes, these biases also corresponded to similar biases in the same regions produced by unseen movements of the participants' own body parts. Furthermore, Oosterhof et al. (2010) found, using MVPA, that classifiers trained on the patterns of response in this general region (EBA was not localized) produced by different observed actions could discriminate among trials in which participants performed those movements out of view (and vice versa). This finding held for both intransitive and transitive actions, and suggests a crossmodal population coding of action in this general region.

Finally, recent evidence has identified a further occipitotemporal region that is relevant to the findings reviewed here. Bracci, Ietswaart, Peelen, and Cavina-Pratesi (2010) reported a hand-selective area that is adjacent to (and dissociable from) left EBA. This area responded selectively to images of hands, relative to feet and other human body parts. Given that they all used hand movements, many of the studies reviewed in this section, including those of Astafiev et al. (2004), Kontaris et al. (2009), Oosterhof et al. (2010), and Peelen and Downing (2005b), merit re-examination with localization of this hand-selective region.

EBA AND FBA CREATE COGNITIVELY UNELABORATED, PERCEPTUAL BODY REPRESENTATIONS

In sum, when stepping back for the broadest view, the above evidence can be characterized by two propositions: (1) EBA and FBA are involved in a wide range of perceptual tasks that involve some aspect of perceiving the human body or its parts; but (2) the evidence for these regions playing a high-level role that is specific to the representation of identity, emotion, body movement, or action, or in the control of motor behavior, is very limited.

We propose instead that the fMRI activations labeled "EBA" and "FBA" reflect relatively concentrated populations of neurons that respond selectively to the visual features that are highly typical of human bodies or body parts. These neurons jointly form a representation of perceived bodies and body parts, making this information available to other regions. They do this by extracting some (but only some) of the information about bodies that is implicit in "early" retinotopic visual representations and making it explicit. This process contributes importantly to computations in other areas in which, for example, representations of emotions, actions, identity, body movements, and goals are extracted.

Coding of visual body/part features by neuronal activity patterns

A reasonable proposal is that this encoding of body information within EBA and FBA is carried out by changing patterns of activity across the neural populations local to these two areas. Taking that approach, we can make specific predictions about the representations formed in EBA and FBA (Figure 4). That is, what aspects of the body are highlighted and made explicit, and what aspects are left implicit?

First, we expect that the patterns of activity in these regions will vary systematically when exposed to different body parts (Op de Beeck, Brants, Baeck, & Wagemans, 2010) perhaps more so in EBA than FBA (Taylor et al., 2007). Likewise, the shape of the body and its parts will be explicitly represented in such a way that even relatively subtle differences in form will be amplified by the neural representation (relative, for example, to representations in earlier visual areas). Furthermore, we expect that the "space" of body shapes will be influenced by experience, so as to be skewed toward commonly experienced body types, sizes, and postures as well as views (Chan, Kravitz, Truong, Arizpe, & Baker, 2010).

Similar arguments can be made for posture —that is, systematically different activity patterns will map to the "space" of different possible postures that bodies can achieve. The coding for posture may be somewhat coarser than for form in light of findings such as those of Urgesi, Candidi et al. (2007). And, as we argued above, some postures may be "special" in that they represent "key frames" that are particularly useful for representing the positions of bodies and limbs efficiently. Note that some studies have examined the response to impossible body postures (Costantini et al., 2005; see also Avikainen, Liuhanen, Schurmann, & Hari, 2003; Cross, Mackie, Wolford, & Hamilton, 2010). We expect these will be coded with reference to the nearest possible posture, perhaps with an additional overall increase in activity due to attentional modulation (Candidi, Urgesi, Ionta, & Aglioti, 2008).

On the other hand, the neurons in EBA and FBA do not explicitly represent many high-level properties. For example, patterns of activity in these regions will relate to body shape rather than identity, so they will be more similar for two bodies or body parts with highly similar shape (even across different viewing conditions such as occlusion, viewpoint, etc.; Taylor et al., 2010) than for

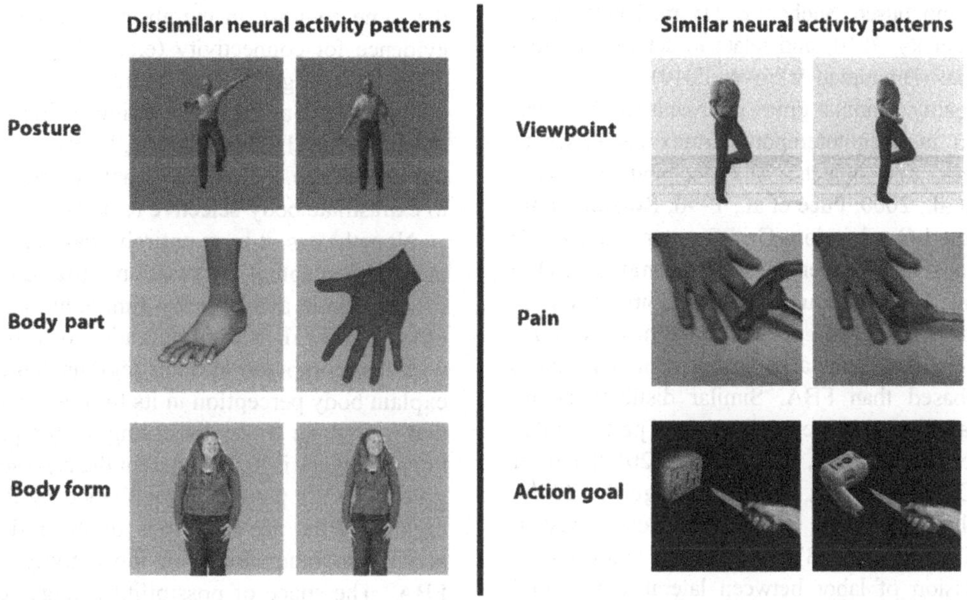

Figure 4. The present framework predicts that, in the extrastriate and fusiform body areas, distinct patterns of activity across neural populations make explicit some aspects of the visual appearance of human bodies but not others. The left panel illustrates the prediction that (other factors being equal), different patterns of neural activity in EBA and FBA will be elicited by seeing the same body holding different postures; different body parts; and different body shapes, sizes, and forms. In contrast, the right panel illustrates the prediction that (other factors being equal), similar patterns of neural activity in EBA and FBA will be elicited by varying viewpoints of the same body (up to some tolerance); the same body part in different emotion/pain contexts; and the same body part undertaking to achieve different goals. Right-side figures adapted from Taylor, Wiggett, and Downing (2010); Morrison et al. (2007); and Bach et al. (2010), respectively.

two images of the same person where shape has changed (e.g., due to weight gain or age), or for two different body parts of the same person. Likewise, where the same body or body part was seen to be moving in different ways, or engaging in different interactions with the same object, neural activity patterns would remain highly similar (assuming other factors were constant). And, again, a similar argument would be made for the case of motor control—the same body stimuli seen in the context of different ongoing motor behavior by the observer would produce similar patterns of activity.

This consideration of patterns of neural activity suggests that neural populations in EBA and FBA are not monolithic but rather may have a complex structure. The most recent evidence has begun to support this idea. For example, as noted above, Bracci et al. (2010) reported a hand-selective region that falls near, but can be dissociated from, EBA; and op de Beeck et al. (2010) identified distinct foci that respond maximally to torsos and to hands. Likewise, Weiner and Grill-Spector (2011) showed evidence for discrete subregions of EBA that fall in a consistent anatomical relationship to hMT+ and retinotopic regions. In the fusiform gyrus, Michels et al. (2009) identified separable subregions that respond differentially to point-light "walker" stimuli depending on the direction in which they are seen to walk. These findings may generally be seen as extensions of the idea of "EBA" or "FBA" that we have developed. That is, they shed light on the local structure of the body representations in these regions, but they do not suggest an entirely different functional role for them.

Local anatomical context

As we and others have discussed (Minnebusch & Daum, 2009; Peelen & Downing, 2007a; Slaughter & Heron, 2004; Taylor et al., 2007), there may be useful parallels to be drawn between the cortical representations of bodies and of other stimulus kinds. The model we develop here can be readily compared, for example, to the structural encoding component of Bruce and Young's (1986) model of face perception, in which the visual properties of faces are extracted and made available to other systems for the analysis of properties such as identification and emotion recognition. This comparison may indeed be more than just an analogy. Bodies and faces, of course, are commonly perceived together, and the close overlap of FFA and FBA hints at functional interactions. Such interactions are further suggested by recent findings of cross-adaptation between faces and bodies (Ghuman, McDaniel, & Martin, 2010) and the discovery that behavioral costs of body inversion depend on the presence of the head (Minnebusch, Suchan, & Daum, 2009; Yovel, Pelc, & Lubetzky, 2010) and relate to activity in face-selective areas (Brandman & Yovel, 2010).

More broadly, a similar gross, neuroanatomical structure is seen in occipitotemporal cortex responses to bodies, faces, and objects: The occipital face area (Gauthier et al., 2000; Puce et al., 1996; Rossion et al., 2003) and the LO subregion (Grill-Spector et al., 1999) of the object-selective lateral occipital complex (LOC) (Malach et al., 1995) are found near EBA. Similarly, FFA and the pFs subregion of LOC are found near FBA. As noted above, EBA forms a body representation that is more part-based than FBA. Similar distinctions are found for faces (stronger representation of parts in OFA than FFA; e.g., Liu, Harris, & Kanwisher, 2010) and for objects (LO vs. pFs; e.g., Drucker & Aguirre, 2009). These parallels and others (e.g., Schwarzlose, Swisher, Dang, & Kanwisher, 2008) suggest there is a domain-general division of labor between lateral and ventral occipitotemporal extrastriate areas. This proposal (and an extensive review of the relevant evidence) is considered in depth elsewhere (Taylor & Downing, 2011).

Recent evidence for broad, diffuse body representations in occipitotemporal cortex may seem at odds with a focus on selective regions such as EBA and FBA. Orlov et al. (2010) found a widespread mosaic of broad regions that respond in subtle but systematic ways to perceiving and moving different body parts (e.g., to upper limbs vs. torsos). These responses encompass (but extend well beyond) EBA and FBA, and these findings suggest that some aspects of the perception (and even movement) of body parts could be represented in a distributed way across this broad cortical territory. These findings do not necessarily challenge the importance of focal, selective regions such as those we examine here. Rather they emphasize that aspects of body perception may extend further than previously thought. There may be multiple representations of bodies (and other stimuli) that coexist at different spatial scales in these areas (cf. Graziano & Aflalo, 2007). Indeed, positing a broad, diffuse occipitotemporal representation of bodies, surrounding local "peaks," may help to account for the familial relationship among many of the distinct but neighboring focal activations discussed here, such as those responding to bodies, faces, tools, action, visual motion, hands, and the like.

Global connections

It is a truism of cognitive neuroscience that task performance is supported by the activity of neural networks. Often, unfortunately, the term "network" is used loosely to describe brain regions that are coactivated in fMRI by the same contrast, in the absence of significant direct evidence for connectivity (e.g., Wiggett & Downing, 2008). Although recent studies address this gap in domains such as face processing, using such tools as dynamic causal modeling (e.g., Fairhall et al., 2007), to our knowledge, there is no extant evidence that pertains to extrastriate body-selective regions.

Nonetheless, it is an entirely reasonable hypothesis that the perceptual information extracted in EBA and FBA is made available by functional connections to other regions. It is the integration of this information with that computed in other regions that together will explain body perception in its fullness. A key question that is at stake when discussing such a putative functional network is to what extent the representations in a given area are penetrated or shaped by those in other regions. What are the limits of the influence of the activity in other areas on the activity in EBA and FBA? The space of possibilities ranges between two extremes. On the one hand, these regions could be entirely encapsulated in the sense that they strictly analyze the lower-level visual signals they receive, and the result is passed on to other brain areas without any further influence by the computations made in those areas. On the other hand, it could be that among the "network" of areas that concerns itself with the perception (and indeed movement) of bodies, the range of functions is widely distributed, and the relatively greater response in one area to visual stimuli and another to motor behavior is a difference of degree rather than kind. On this view, in the extreme it makes little sense at all to consider EBA (or FBA) in isolation, as its functions can only be understood in the context of the entire network.

Our speculation is that the truth leans closer to the former picture than the latter. There are good functional arguments for why the perceptual functions of a network should not be fully penetrable by the operations of other brain areas; a relatively stimulus-bound, veridical representation can be important, for example, when expectations are violated. An important exception to this principle is that the representations in body-selective brain areas (and other extrastriate areas) will be strongly modulated by the focus of attention. This may occur indirectly, such as by spatial attention mechanisms affecting retinotopic maps, with the knock-on effect of modulating body representations according to the contents of attended locations (cf. Downing, Liu, & Kanwisher, 2001). Additionally, attention may directly modulate activity in body-selective regions as a way of supporting a template in the service of an abstractly specified task (Peelen et al., 2009). Given some of the considerations above about substructure, we may further suppose that attentional effects can be general—across whole body-selective areas—or target specific limbs, say, where those are most relevant.

On our account, a body-selective neuron in EBA or FBA does not "know" (for example) whether someone is angry or sad—it does not contain "angry body detectors." There is no systematic mapping between the patterns of neural activity in these regions, on the one hand, and the body postures related to different emotions, on the other hand. Other regions interpret the emotion of seen bodies, on the basis of the information provided by EBA and FBA as well as other areas that represent semantic knowledge, scene context, and the like. One result of this process may be attentional feedback to occipitotemporal cortex. If the hands, for example, are important for expressing emotions, then attention will tend to select the hands in the image when emotion is relevant. Hand-related activity patterns in the body-selective areas will be enhanced accordingly. In this way, attention modulates the responses of body-selective neurons to ensure that resources are allocated efficiently to process behaviorally relevant input.

Puzzles

We have focused on vision, but some new evidence points to occipitotemporal body representations in other modalities. Kitada, Johnsrude, Kochiyama, and Lederman (2009) found separate subclusters in the lateral occipitotemporal cortex that respond selectively to either visually or haptically presented body parts. More recently, Costantini et al. (2011) found evidence for right (but not left) EBA engagement by haptic aspects of body-part perception (compared to non-body objects). When considered alongside studies of object recognition that show an overlap between tactile and visual representations (Amedi, von Kriegstein, Van Atteveldt, Beauchamp, & Naumer, 2005), and "typical" occipitotemporal patterns of selectivity in blind participants for objects (Mahon, Anzellotti, Schwarzbach, Zampini, & Caramazza, 2009; Pietrini et al., 2004) and actions (Ricciardi et al., 2009), these findings point to nonvisual influences on the functional organization of the extrastriate cortex. With respect to the present discussion of EBA and FBA, however, we predict that where tactile stimulation engages these regions, it will be in the service of similar functional properties to those discussed here for the domain of vision.

Some further puzzles remain in the poorly understood influences of nonsensory factors on the characteristics of EBA and FBA. For example, the handedness of participants is related to the lateralization of EBA (but not hMT+) and FFA, such that right-handers, but not left-handers, show a right-hemisphere bias in the size of these selective regions (Willems, Peelen, & Hagoort, 2010). In parallel, it appears that women show a stronger right-lateralization of EBA and FBA than men (in a study in which all bar one of the participants were right-handers; Aleong & Paus, 2010). This may relate to left-visual-field specific body size estimation deficits found in women but not men (Mohr, Porter, & Benton, 2007). The present framework does not account for these (possibly interacting) effects of handedness, gender, and hemisphere.

CONCLUSION

We have sought to take a deliberately simple account of occipitotemporal body representations as far as possible, and, in doing so, to flush out gaps in our knowledge and to provoke debate. This simple model succeeds well in accounting for extant data. Many findings that appear to show elaborated body or person representations in EBA or FBA can be explained with an appeal to simpler principles such as the influence of attention. We argue that body-selective regions exist because of the important and complex information that body parts convey. The fMRI activations that we label "EBA" and "FBA" reflect the activity of neurons that analyze the form and configuration of bodies in a highly efficient and specialized manner. This information is then made available to other brain areas that extract and make explicit the meaningful signals that are implicit in the visual appearance of the body. The joint activity of these regions is what constitutes understanding of an image; but, we argue, the influence of activity in other regions on EBA and FBA is largely limited to attentional modulation.

This discussion helps to highlight some of the many challenges for future research. In our view, these include the following: discovering the detailed structure of body representations in EBA and FBA and how they may differ from each other; investigating their similarities and differences to anatomically aligned object and face representations; obtaining further evidence on the nonvisual influences on activity in these regions; understanding how they relate to diffuse patterns of visual and motor body-related information distributed throughout occipitotemporal cortex; and testing explicitly the proposal that these regions form functional connections with other brain areas in order to jointly create coherent, contextualized, and useful representations of the body.

Commentaries

Do body-part concepts depend on the EBA/FBA?

David Kemmerer[1,2,3]
[1]Department of Speech, Language, and Hearing Sciences, Purdue University, West Lafayette, IN, USA
[2]Department of Psychological Sciences, Purdue University, West Lafayette, IN, USA
[3]Department of Neurology, Division of Cognitive Neuroscience, University of Iowa College of Medicine, Iowa City, IA, USA
E-mail: kemmerer@purdue.edu

http://dx.doi.org/10.1080/17588928.2011.604718

Abstract: Downing and Peelen argue that the EBA/FBA represent body-part shapes in a highly schematic manner that is independent of personal identity, emotional expression, movement pattern, and action goal, and that cuts across visual and haptic modalities. According to the grounded cognition framework, these properties make the EBA/FBA suitable for processing body-part shapes not only for perceptual purposes but also for conceptual purposes. Any account of the neural substrates of body-part concepts must, however, accommodate significant cross-linguistic diversity in this semantic domain. Hence, an alternative possibility is that the shape components of body-part concepts depend on areas adjacent to the EBA/FBA.

A prominent theory of conceptual knowledge is the grounded cognition framework, which maintains that concepts are not completely divorced from modality-specific systems for perception and action, but instead overlap with them to some degree (Barsalou, 2008; Kiefer & Pulvermüller, 2011; Martin, 2007). Consider, for example, an fMRI study by Chao, Haxby, and Martin (1999) that focused on the perceptual and conceptual processing of animals and tools. Perceptual processing was probed with passive viewing tasks and match-to-sample tasks, and conceptual processing was probed with picture-naming tasks and property-verification tasks. Across all of the tasks, perceptual as well as conceptual, greater activation for animals was found in a lateral portion of the mid-fusiform gyrus, whereas greater activation for tools was found in a medial portion of the mid-fusiform gyrus. These regions probably represent the shapes of animals and tools, and the fact that they were activated not only by pictures, but also by words, supports the grounded cognition framework. Moreover, several other studies have generated convergent results which suggest that the word-induced fusiform activations reflect the retrieval of shape properties for conceptual purposes, and are not due to mere "imagery" (Martin, 2007).

Although much has been learned about the neural substrates of the shape properties of animal and tool concepts, almost nothing is known about the neural substrates of the shape properties of body-part concepts, as expressed by terms like *arm, hand, leg,* and *foot.* This gap in the literature is surprising, given the tremendous interest in body representation that has recently emerged. There is some neuropsychological evidence that body-part concepts depend on the lateral/ventral occipitotemporal cortices, among other regions (Kemmerer & Tranel, 2008). To my knowledge, however, no studies have tested the hypothesis, derived from the strongest form of the grounded cognition framework, that the shape properties of body-part concepts rely specifically on the EBA/FBA or on neighboring areas that also seem to contribute to body-part perception.

According to Downing and Peelen, the EBA/FBA are tuned to the domain of body-part shapes in several ways: They capture "bare bones" schematizations of these objects, as shown by their sensitivity to silhouettes, cartoons, and stick figures; they respond to body-part shapes regardless of personal identity, emotional expression, movement pattern, and action goal; and they generalize across visual and haptic modalities. Interestingly, these representational capacities appear to make the EBA/FBA well suited to processing body-part shapes for both perceptual and conceptual purposes. Hence, an advocate of the grounded cognition framework might predict that understanding a word like *leg* involves activating the same patches of the EBA/FBA, or the same population codes in these regions, that are engaged during the perceptual recognition of legs.

It is noteworthy, however, that although most languages have words that segment the body according to salient perceptual discontinuities, there is still substantial variation. To take a few examples, Savosavo has a single category for leg that begins at the hip joint and encompasses the foot, ignoring both knee and ankle discontinuities; Yélî Dnye has one term for the upper leg and another term for the lower leg plus the foot, thus respecting the knee discontinuity but ignoring the ankle discontinuity; Jahai recognizes all three parts—upper leg, lower leg, and foot; and Tidore has one term that covers the foot, lower leg, and lower part of the thigh, and a separate term for "groin" that includes the upper part of the thigh, thus making a category distinction in the absence of a corporeal discontinuity (Majid, 2010).

Any account of the neural underpinnings of body-part concepts must accommodate this diversity. Perhaps language is one of the experiential/cultural factors that, according to Downing and Peelen, influences the EBA/FBA. Alternatively, there may be multiple occipitotemporal body-part maps, and the shape properties of body-part concepts may be subserved by areas near the EBA/FBA. In this situation, the relationships between perceptual and conceptual body-part representations would probably be looser, but they would still be quite close, consistent with a weaker form of the grounded cognition framework. The key point is that both of these possibilities, and others, are ripe for investigation.

* * *

Adaptation studies suggest interactive feedback shapes responses in occipitotemporal regions

Michael P. Ewbank

MRC Cognition and Brain Sciences Unit, Cambridge, UK

E-mail: michael.ewbank@mrc-cbu.cam.ac.uk

http://dx.doi.org/10.1080/17588928.2011.604719

Abstract: Downing and Peelen portray the EBA and FBA as a largely impenetrable system, passing on visual signals to other cortical areas where high-level information is extracted. They suggest that the influence of other regions on EBA and FBA is primarily limited to changes in attentional modulation. In the following commentary, this proposal is evaluated within the context of a predictive coding framework. Recent evidence is discussed indicating that fMRI-adaptation effects in occipitotemporal cortex, including EBA and FBA, are consistent with the influence of higher-level modulation, suggesting that responses in these regions are shaped through the interactive feedback of a hierarchical network.

In their discussion paper, Downing and Peelen propose that the joint activity of the extrastriate body area (EBA) and the fusiform body area (FBA) creates a detailed representation of the visual appearance of the human body. They also suggest that the perceptual functions of these regions are largely impenetrable by the operations of other brain areas. By their own admission, the authors present an oversimplified account of body representations in the occipitotemporal cortex. However, a simple model need not characterize EBA and FBA as cognitively impenetrable modules; instead, findings from fMRI-adaptation studies suggest that responses in these regions can be shaped through interactive feedback.

Recently, we measured fMRI-adaptation in EBA and FBA to presentations of the same body compared to different bodies (Ewbank et al., 2011). Both regions showed adaptation to bodies that was invariant to substantial changes in size and view. One interpretation of this finding is that both EBA and FBA hold a relatively high-level representation of body identity. However, using dynamic causal modeling, we were able to demonstrate that adaptation is associated with a change in effective connectivity from FBA to EBA, indicating that adaptation in EBA—across changes in size and view—is the consequence of modulation from FBA. Our findings thus provide direct evidence that top-down modulation may underlie visual adaptation within the occipitotemporal cortex. Moreover, other evidence suggests that such occipitotemporal adaptation effects can be modulated by top-down input from other "systems." For example, Summerfield et al. (2008) found greater adaptation to faces in the fusiform face area (FFA) when repetitions were expected, reflecting top-down influences from the prefrontal cortex.

These effects are consistent with models of predictive coding in which adaptation is thought to reflect a decrease in prediction error between bottom-up (stimulus-related) and top-down (prediction-related) inputs (Friston, 2005;

Henson, 2003). Indeed, in a previous study, Downing and colleagues themselves suggested that adaptation in EBA and FBA may reflect higher-level modulation (Taylor, Wiggett, & Downing, 2010). Here, they found that the presentation of scrambled masks disrupted adaptation to images of the same body across view, and suggested that this effect may reflect an interruption of the later stages of neural activity in EBA and FBA, "which may be influenced by reciprocal interactions with higher level cortical areas." ERP evidence also indicates that adaptation across view occurs in a later time period than adaptation to identical images of objects (Schendan & Kutas, 2003). These results suggest that top-down influences on EBA and FBA are unlikely to be restricted to attentional modulation.

In the current paper, the authors propose that a "stimulus-bound" representation in EBA and FBA "may be important for determining when expectations are violated." This suggests neurons in these regions respond monotonically to all input (unless modulated by changes in attention and other factors). The advantage of a hierarchical model, however, is that it enables the brain to construct prior expectations in a dynamic fashion—rather than being stimulus bound, energy-costly signals are focused on rare inputs, alerting the system to important changes. Thus, neurons in EBA and FBA are suppressed when the same body is expected (e.g., during repetition)— including transitions in size and view—but remain sensitive to unpredicted transitions, such as random changes in body shape (Aleong & Paus, 2010). Similarly, EBA and FBA show an increased response to contorted compared to normal body postures (Cross, Mackie, Wolford, & Hamilton, 2010) and incoherent compared to coherent action sequences (Downing, Peelen, Wiggett, & Tew, 2006).

The key aspect of predictive coding models is that each level in a hierarchy is penetrable by predictions received from the level above. The FBA is unlikely to correspond to the highest stage of body processing, and therefore responses in this region are also likely to be modulated by inferences in higher-level areas. Within this framework, adaptation to repetition of the same body is the consequence of interactions between regions rather than being localized to an individual area.

It appears likely that EBA and FBA extract visual information that contributes to computations pertaining to identity. However, adaptation studies suggest that the relationship between EBA and FBA, and the relationship between these regions and higher-level areas, is likely to be reciprocal, with activity in higher-levels areas dynamically shaping responsivity in lower-levels regions.

* * *

Human body perception and higher-level person perception are dissociated in early development

Virginia Slaughter

Early Cognitive Development Centre, School of Psychology, University of Queensland, Brisbane, Australia

E-mail: vps@psy.uq.edu.au

http://dx.doi.org/10.1080/17588928.2011.604720

Abstract: Developmental data support the proposal that human body perceptual processing is distinct from other aspects of person perception. Infants are sensitive to human bodily motion and attribute goals to human arm movements before they demonstrate recognition of human body structure. The developmental data suggest the possibility of bidirectional linkages between EBA- and FBA-mediated representations and these higher-level elements of person perception.

Downing and Peelen's conclusion that EBA and FBA encode structural details of the human body and its parts, but do not directly contribute to higher-level person perception, is consistent with developmental work on these topics. Behavioral research with human infants reveals dissociations between body perception and many of the other capacities discussed by Downing and Peelen. I will focus on two of them: recognition of human motion patterns and detection of goal-directed human movement.

The available evidence indicates that EBA and FBA play a causal role in adults' visual perception of the human body. This is evident in differential responsiveness of these regions to intact human body images as compared to scrambled bodies or other objects. In our behavioral studies with infants, we evaluate visual discrimination of intact human bodies from scrambled non-bodies constructed by moving the arms and legs from their canonical locations. This task indicates that reliable recognition of human body structure does not emerge until the end of the first year of life (Heron & Slaughter, 2010). We have speculated that this is because, prior to 9–12 months, infants have little visual access to whole bodies; they regularly see other people's faces, their own and others' hands (Aslin, 2009), and their own legs. But they may not have much visual experience of how these parts fit together. So prior to the end of the first year of life, infants do not appear to have access to detailed structural representations of the human body.

By contrast, perception and recognition of human body movement emerges within the first few months of life. Behavioral studies indicate discrimination of intact versus phase-shifted human point light display (PLD) walkers by 3 months of age and discrimination of upright from inverted PLD walkers by 5 months (see Bertenthal, 1993). Recent ERP experiments with 8-month-olds revealed different patterns of parietal activation while infants view normal versus scrambled human PLD walkers (Hirai & Hiraki, 2005) as well as upright versus inverted ones (Reid, Hoehl, & Striano, 2006). Therefore, perception of whole-body human movement patterns is an earlier achievement than perception of human body structure.

Attributing goals to human movement is also developmentally distinct from body perception. By 6 months of age, infants encode the goal of a moving human arm (Kiraly, Jovanovic, Prinz, Aschersleben, & Gergely, 2003). That is, when they view an arm reaching for one of two objects, infants expect the arm's next reach to go for the same object, consistent with its prior goal. Little is currently known about infants' perception of individual, salient body parts. However, a recent study revealed that 6-month-olds attribute goals to human movement even when it involves a violation of body structure (e.g., when the arm moving to grasp an object bends impossibly) (Southgate, Johnson, & Csibra, 2008). So goal attribution to human reaching movements emerges prior to whole-body recognition, and also appears to be distinct from expectations about how arms themselves should appear.

Downing and Peelen favor a model in which EBA- and FBA-mediated structural representations of the body are functionally encapsulated from higher-level inferential systems. They imply that perceptual information computed by body-selective areas constitute representational primitives that are then passed "upward" for more complex processing. This is a logical model that may well describe what occurs in the adult brain. However, as described above, some aspects of higher-level processing (e.g., of human body movement and goal-directed actions) are evident before infants exhibit knowledge about human body structure. This means that at least some aspects of higher-level person perception can occur without the kind of information computed by EBA and FBA. It also suggests that body structural representations may be derived from (rather than be a precursor to) those higher-level processes, or that there may be bidirectional linkages, at least early in development.

* * *

No two are the same: Body shape *is* part of identifying others

Richard Ramsey[1,2], Hein T. van Schie[3], and Emily S. Cross[2,3]

[1]Faculté de Psychologie et des Sciences de l'Education, Université Catholique de Louvain, Louvain-la-Neuve, Belgium

[2]Wales Institute of Cognitive Neuroscience, School of Psychology, Bangor University, Bangor, Gwynedd, UK

[3]Behavioural Science Institute, Radboud University, Nijmegen, The Netherlands

E-mail: richard.ramsey@uclouvain.be

http://dx.doi.org/10.1080/17588928.2011.604721

Abstract: Downing and Peelen argue for a clear distinction between body and identity representation, with the former performed by EBA and FBA, and the latter performed elsewhere in the brain. Under a predictive coding account, we argue that this separation is unnecessary: Representing bodies is part of representing identity. While neurons in EBA and FBA may only code for body shape and posture, we propose that they are a part of a reciprocally connected cortical network that functions to minimize prediction error when making identity inferences. We propose a novel way to test the hypothesis that EBA and FBA are critically involved in person identification.

Downing & Peelen offer a "cognitively unelaborate" interpretation of research examining EBA and FBA. The authors argue that "the role of these regions [EBA and FBA] consists in creating a perceptual representation of the shape and posture of the body and its parts, which may then be used by other brain regions (e.g., the anterior temporal lobes; Kriegeskorte et al., 2007) to represent person identity explicitly" (p. 193).

Such a proposal assumes that the coding of physical features is separate from a process that determines identity. However, the authors allow for exchange between these distinct processes (and brain regions). For example, "Note that our account does not exclude the possibility that responses in EBA and FBA can be influenced by identity through top-down modulation: The sight of a romantic partner (or, indeed, the self) may increase responses in EBA/FBA due to increases in attention and arousal" (p. 193). Under this framework, it appears that body processing in extrastriate cortex is *not* part of a "who" system involved in person identification (Georgieff & Jeannerod, 1998), but rather supplies input to it by passing on body shape and postural information. This implies that body shape and posture are not part of "higher-level" identity processes. Rather, EBA and FBA neurons code physical features in an agent-blind manner; that is, they do not contribute to differentiating between identities.

While we agree that current evidence does not support an interpretation beyond coding of physical features, we propose that EBA and FBA *do* play a critical role in understanding identity by being part of a network of reciprocally connected neural regions that "bias" the neural signals involved in making identity inferences. To illustrate this point, we revisit the "romantic partner" example (p. 193). When a romantic partner walks into a room, coding of her physical features in EBA and FBA, such as a small head and skinny arms, would bias a "who" system toward supporting an inference about the person being a romantic partner, rather than a different acquaintance (with a distinct body shape). This biasing process is similar to the predictive coding account of action perception, which hypothesizes that the brain relies upon Bayesian models to predict how an action should unfold across time, based on prior experience (Kilner, Friston, & Frith, 2007). Perceiving a body shape that matches your romantic partner's does not mean it is *definitely* your romantic partner, but based on your prior experience with the perceived body shape, it is statistically more likely to be your partner than another person. In other words, while neurons in EBA and FBA may code only body shape and posture, in doing so they contribute to reducing prediction error throughout a reciprocally connected network of brain regions, which together determine one's current identity inference. Importantly, EBA and FBA generate only part of the biasing signal that contributes to making an identity inference; one must link this signal with additional biasing signals that relate to other prior person experiences, such as accent, gait, clothes, hairstyle, etc., and this, we agree, likely occurs elsewhere in this cortical network, beyond EBA and FBA.

To test whether EBA and FBA are involved in identifying "who" somebody is according to a predictive coding framework, we suggest that standard neuroimaging experiments lack the sensitivity to address this issue (e.g., Hodzic, Muckli, Singer, & Stirn, 2009); instead, we advocate the use of causality mapping techniques, such as dynamic causal modeling and Granger causality mapping. In a neuroimaging experiment where participants must identify familiar and unfamiliar bodies, causality mapping would enable the flow of information processing to be traced from occipitotemporal areas to anterior regions (e.g., anterior temporal lobes), and the inverse direction. If EBA and FBA do indeed play a role in identifying "who" somebody is, differential directional flow between these regions should emerge when identifying familiar versus unfamiliar bodies. For example, in perceiving familiar individuals, there may be more information flow "backward" from anterior to posterior brain regions, reflecting a top-down biasing signal based on prior information about the person. In contrast, in perceiving unfamiliar individuals, there may be more information flow "forward" from posterior to anterior brain regions, reflecting more bottom-up processing of body shape and posture.

In summary, Downing and Peelen imply that body shape and posture are coded independently of identity, as identity only involves what the authors label as "higher-level processes." Whether they are labeled high- or low-level, we argue that unique features of a person's body contribute to knowing who a person is. Moreover, we suggest that EBA and FBA are part of a cortical network of brain areas that rely upon Bayesian models to predict likely identity from prior person experience. As such, we argue that EBA and FBA are integral nodes of a "who" system for making identity inferences. We suggest that a worthwhile pursuit at this stage would be to employ more sensitive neuroimaging measures to determine whether EBA and FBA might assist with body identification according to a Bayesian framework of predictive coding.

* * *

When perception and attention collide: Neural processing in EBA and FBA

Susanne Quadflieg and Bruno Rossion
Department of Psychology, Catholic University of Louvain, Louvain-La-Neuve, Belgium
E-mail: susanne.quadflieg@uclouvain.be

http://dx.doi.org/10.1080/17588928.2011.604722

Abstract: In their timely review, Downing and Peelen differentiate between neural systems dedicated to body perception and body-based person inferences. Rather than supporting their view of these operations as largely separate mental and neural entities, we emphasize reasons to favor a stronger notion of co-dependency.

Based on their discussion of research findings concerning the extrastriate body area (EBA) and the fusiform body area (FBA), the authors conclude that these regions provide a "cognitively unelaborated perceptual framework [of bodies] that allows other cortical systems to exploit the rich, socially relevant information." In so doing, the authors iterate the idea of distinguishing between a core neural system of person perception and an extended system of person inferences as originally introduced by Haxby, Hoffman, and Gobbini (2000) in the realm of face processing. According to this approach, one set of brain regions analyzes the visual appearance of people, whereas additional sites subsequently use this information to generate person inferences (e.g., regarding a target's identity, see Macrae & Quadflieg, 2010).

An extensive body of work indicates, however, that connectivity between brain regions is usually bidirectional and that information reaching a specific cortical site tends to be processed under the influence of cross-regional interactions (e.g., Friston, 2005). From a computational point of view, it also seems unlikely that feed-forward mechanisms alone can accomplish the rapid and adequate recognition of complex visual input (e.g., Mumford, 1992). Hence, it has been argued that perceptual processes are habitually penetrated and disambiguated by higher-level cognitive factors such as a perceiver's expectations and processing goals (Bar, 2009; Bruner, 1973).

To counter such a view in the realm of body perception, Downing and Peelen emphasize that regions such as the EBA and FBA achieve "relatively stimulus-bound" representations that are "not fully penetrable by the operations of other brain areas" (p. 202). Alas, the authors own choice of words gives away the caveat. How *relatively* stimulus-bound are the representations and to what extent (if not *fully*) are they modifiable? Importantly, accepting the idea of penetration does not result in the assumption that EBA and FBA themselves compute higher-order cognitive inferences—in which case, there would be no necessity for penetration. Rather, the assumption of penetration argues that due to operations in other brain sites, processes unfolding in the target region are modified.

To illustrate this idea, consider the following results (Quadflieg et al., 2011): Perceivers asked to sex-categorize men and women portrayed in various types of occupational clothing displayed enhanced activity in EBA and FBA when a target's attire violated common gender stereotypes (e.g., when seeing a female firefighter). These data seem incompatible with Downing and Peelen's claim that core body-perception areas create cognitively unelaborated representations. Of course, EBA and FBA are not argued to "know" whether a target fits cultural stereotypes. Nevertheless, the finding indicates that "higher-order" expectations can help or hinder building a coherent person percept.

Note that changes in EBA and FBA activation across experimental conditions as illustrated here may reflect mere attentional modulations during stimulus processing. According to Downing and Peelen, such modulations may be caused by sampling differences (i.e., differences in preferentially attended-to contents) or template differences (i.e., differences in body templates used in the service of task completion). Intriguingly, both mechanisms are also listed by the authors to characterize the type of perceptual processing occurring in core body-perception regions. Populations of neurons in EBA and FBA are argued to extract and bind body information in a non-exhaustive manner (only *some* of the information) and based on templates "useful for representing the positions of bodies and limbs" (p. 200). This observation then raises the question of how attention-related modulations actually differ from "perceptual" modulations if both comprise the selective sampling, binding, and template-based encoding of body information.

In our view, an empirically falsifiable differentiation between perceptual and attentional modulations forms a necessary prerequisite for understanding aspects of penetration in body perception. This lofty goal cannot be addressed by fMRI alone. Rather, dissecting the exact temporal unfolding of body processing across brain regions through high-density recordings of EEG/MEG in humans (e.g., Appelbaum, Wade, Vildavski, Pettet, & Norcia, 2006), or through direct recordings of single-cell responses in nonhuman primates (e.g., Pinsk et al., 2009), is likely to help with such a feat. Alternatively, investigating whether receptive field properties of neurons located in EBA and FBA vary depending on expectations and task demands (see David, Hayden, Mazer, & Gallant, 2008) will add to our understanding of whether it is person perception that solely informs person inferences, or whether person inferences also shape person perception.

* * *

Differential contributions of occipitotemporal regions to person perception

Annie W.-Y. Chan and Chris I. Baker
Laboratory of Brain and Cognition, National Institute of Mental Health, NIH, Bethesda, MD, USA
E-mail: chanannie@mail.nih.gov

http://dx.doi.org/10.1080/17588928.2011.604723

Abstract: Downing and Peelen have produced an excellent review synthesizing the current literature on the processing of body stimuli in visual cortex. However, while they consider the extrastriate body area (EBA) and fusiform body area (FBA) together, these regions are physically separate in cortex and likely contribute differentially to person perception. Here, we evaluate the hierarchical view of processing in EBA and FBA and highlight the visual field biases in these regions, which may provide insight into their origin and functional roles.

We agree with Downing and Peelen that representations in EBA and FBA primarily reflect visual features of bodies, not higher-level cognitive information such as identity or emotion. However, there are two critical aspects of these regions that remain unclear; namely, why are there two regions and what are the differential contributions of EBA and FBA to person perception?

Are EBA and FBA hierarchical?

Downing and colleagues have proposed that EBA primarily represents individual body parts whereas FBA represents the whole body (combination of body parts). Support for this view comes primarily from a study (Taylor, Wiggett, & Downing, 2007) demonstrating that, while the selectivity in EBA increased gradually with increasing amounts of the body presented (finger, hand, limb, whole body), FBA exhibited a step-like function with no selectivity for single fingers or hands. Such a hierarchical account is similar to that proposed for faces and objects, for which there are also two selective regions, one lateral and one ventral (e.g., Liu, Harris, & Kanwisher, 2010). However, as pointed out elsewhere (Op de Beeck, Brants, Baeck, & Wagemans, 2010), Taylor and colleagues did not counterbalance the body stimuli such that each body part occurred in both isolated and combined conditions. Thus, for example, torsos only ever occurred in combination with fingers, hands, and arms, and apparent selectivity for combinations of body parts could be confused with torso selectivity, which has been reported near FBA (Op de Beeck et al., 2010; Orlov, Makin, & Zohary, 2010). Further, other studies have found no difference in selectivity in FBA for isolated, mixed body parts compared with headless bodies (Schwarzlose, Baker, & Kanwisher, 2005). Thus, the current evidence for hierarchical processing within body selective regions is not entirely compelling.

What do visual field biases tell us about EBA and FBA?

Both EBA and FBA show spatial biases within the visual field, and this may be important for understanding their roles in person perception. First, FBA lies in a region of cortex with a foveal bias (Levy, Hasson, Avidan, Hendler, & Malach, 2001) and exhibits stronger responses to foveal than peripheral stimuli (Schwarzlose, Swisher, Dang, & Kanwisher, 2008). In contrast, EBA shows stronger responses for peripheral than foveal stimuli (Schwarzlose et al., 2008), and much of EBA appears to overlap regions with an underlying preference for peripheral space (Weiner &

Grill-Spector, 2011). Thus, while FBA appears to be foveally biased, EBA appears to have a bias for more peripheral parts of the visual field.

Second, EBA shows an elevation bias with stronger responses for lower compared to upper visual field body stimuli (Schwarzlose et al., 2008), and much of EBA overlaps maps of the lower visual field (Weiner & Grill-Spector, 2011). This lower-field bias is also observed in nearby regions selective for motion (middle temporal area, MT; Maunsell & Van Essen, 1987) and objects (Kravitz, Kriegeskorte, & Baker, 2010), and is consistent with their relative proximity to the lower field representation in early visual cortex. In contrast, FBA does not show a clear bias to either the upper or the lower visual field (Schwarzlose et al., 2008), perhaps because of the strong foveal bias, although an upper-field bias has been reported in nearby regions of ventral occipitotemporal cortex, such as object-selective posterior fusiform sulcus (Kravitz et al., 2010). Body parts primarily occur outside the fovea in the lower visual field, and this may account for the larger size of EBA relative to FBA (Chan, Kravitz, Truong, Arizpe, & Baker, 2010). In addition, a lower-field bias has also been reported in regions of the dorsal visual pathway involved in the visual control of action (Danckert & Goodale, 2003).

In sum, while the specific roles of EBA and FBA in person perception are unclear, the underlying visual field biases in these regions must be taken into account in any theory of their relative functional roles. These underlying visual field biases may reflect the developmental and evolutionary origin of EBA and FBA. More data are needed to test the hierarchical account and clarify the nature of the underlying visual field biases.

* * *

The extrastriate body area (EBA): One structure, multiple functions?

Floris P. de Lange and Harold Bekkering
Radboud University Nijmegen, Donders Institute for Brain, Cognition, and Behavior, Nijmegen, The Netherlands
E-mail: floris.delange@donders.ru.nl

http://dx.doi.org/10.1080/17588928.2011.604724

Abstract: Downing and Peelen argue that the extrastriate body area (EBA) creates an unelaborated visual representation of the human body, but is not implicated in any higher-order computational process. We believe that this reflects an outdated view of brain function, in which neural regions are informationally encapsulated modules with fixed computational properties. In contrast, there is mounting evidence that functional properties may not be fixed but may be dependent on the context in which the region is recruited. We will illustrate this by taking a closer look at a visual area of which the properties are potentially even simpler: the primary visual cortex.

The ultimate goal of science is to reduce complexity. In this regard, Downing and Peelen do science a service, by "taking a deliberately simple account of occipitotemporal body representations as far as possible" (p. 203). The claim of Downing and Peelen is that EBA and FBA are not involved in any high-level functions such as perceiving body identity or action goals, but only encode fine details about visually perceived bodies. While we do not contest the importance of these regions for visual encoding of perceived bodies, we take issue with the notion that these regions have no functional relevance beyond this. In particular, we feel that this reflects an outdated view of brain function, in which neural computations are time- and context-independent, whereas mounting evidence paints a more dynamic picture of neuronal computations. We will illustrate this with an example from an earlier visual cortical region, the primary visual cortex (V1).

Since the time of Hubel and Wiesel (1968), V1 has been traditionally thought of as an "edge detector" or spatial filter, given that V1 neurons are exquisitely tuned to both the retinotopic location and spatial orientation of a visual stimulus (e.g., Bell & Sejnowski, 1997). Indeed, the initial response of V1 neurons is very reminiscent of an "edge detector," with an exquisitely tuned response of each neuron to a particular orientation in a particular part of visual space. This response pattern has been robustly observed both in awake behaving animals and under profound states of anesthesia (Dow, Snyder, Vautin, & Bauer, 1981; Schiller, Finlay, & Volman, 1976).

It may, however, be a mistake to conclude that V1 is only functionally relevant to edge detection. Subsequent studies have shown that V1 response properties (such as orientation and color tuning) can be dramatically different at *later* time points during perceptual inference (Lamme & Roelfsema, 2000). These changes in response profile, which are likely due to feedback (and possibly horizontal) connections to V1, suggest that V1 computations may be different at different time points. Indeed, computational models of perceptual inference suggest that while early activity in V1 may simply reflect feature detection, at a later stage V1 may actively participate in multiple visual routines such as object reconstruction and completion (Lee, Mumford, Romero, & Lamme, 1998). Indeed, later activity in V1 appears to reflect the perception of illusory contours in Kanisza-like figures, where no physical line is present in the neuron's receptive field (Lee & Nguyen, 2001), and it has even been argued that feedback to V1 is necessary in order to become aware of a visual stimulus (Lamme, 2006; Pascual-Leone & Walsh, 2001). Finally, a recent study provides compelling evidence for a potential role of V1 in working memory, when a visual stimulus had to be held online for a prolonged period of time while the stimulus is not physically present (Harrison & Tong, 2009). Considering all the evidence, the label of V1 as an "edge detector" may at best be an incomplete characterization of this cortical module. While this label may correctly describe the function of this node at an early stage of the cortical computation, when receiving information from the lateral geniculate nucleus in the thalamus, its response properties and thereby putative computational role appear to change at later stages when it receives input from other areas that are higher in the cortical hierarchy.

The same argument may well pertain to the EBA and FBA. There is indeed good evidence of these regions' involvement in the encoding of fine details of bodies (Downing, Jiang, Shuman, & Kanwisher, 2001), just as there is good evidence of the involvement of V1 in encoding orientation and contrast of line stimuli (Hubel & Wiesel, 1968). But there appears equally good evidence of the modulation of activity of these regions by more high-level factors. Indeed, Downing and Peelen describe several cases in which EBA/FBA activity is modulated by higher-level factors. They interpret these activity modulations as "attentional feedback to occipitotemporal cortex" (p. 203). However, as we have seen with the example of V1, activity modulations by feedback may subserve much more complex processing than simply "boosting the representation," and may alter the representation itself. The activity of V1 to an illusory (Lee & Nguyen, 2001) line is "cognitively elaborated," in the sense that it does not passively represent "what the eyes tell," but rather what is in "the mind's eye." It may be equally incorrect to refer to the representation in EBA as "cognitively unelaborated."

Perhaps the most striking examples of top-down influence on EBA are provided by motor control studies, in which no visual stimulus of a body part were presented, yet where robust EBA activity is sometimes observed (Astafiev, Stanley, Shulman, & Corbetta, 2004; Kuhn, Keizer, Rombouts, & Hommel, 2011; Zimmermann, Meulenbroek, & de Lange, 2011). This situation is reminiscent of the aforementioned study by Harrison and Tong (2009), where, in the absence of any physical stimulus, V1 contributes to holding online a perceptual representation. We believe that EBA may have a similar role in motor control, in terms of holding online a visual representation of the goal state of an action. While this does not contest a role of EBA in representing bodies, it does not render the region irrelevant to motor planning. A crucial test of the necessity of this region in action planning should come from intervention studies, such as lesion studies or transcranial magnetic stimulation (TMS). Indeed, a patient study observed deficits in motor planning following lesions of the ventral occipitotemporal cortex (Dijkerman, McIntosh, Schindler, Nijboer, & Milner, 2009), though the relative lack of specificity precludes a straightforward interpretation. Importantly, TMS studies that aim to show the (ir) relevance of EBA to motor control should bear in mind the potential dynamic nature of the computations in EBA, possibly by targeting this area at different stages of the planning process (cf. Pascual-Leone & Walsh, 2001). We believe that this could be a fruitful area of future research.

* * *

Functional and epiphenomenal modulation of neural activity in body-selective visual areas

Cosimo Urgesi[1] and Alessio Avenanti[2]

[1]Dipartimento di Scienze Umane, Università di Udine, Udine, Italy; and Istituto di Ricovero e Cura a Carattere Scientifico "Eugenio Medea", Polo Friuli Venezia Giulia, Pordenone, Italy

[2]Dipartimento di Psicologia, Alma Mater Studiorum – Università di Bologna, Bologna, Italy; and Centro

Studi e Ricerche in Neuroscienze Cognitive, Polo Scientifico-Didattico di Cesena, Alma Mater Studiorum – Università di Bologna, Cesena, Italy
E-mail: cosimo.urgesi@uniud.it

http://dx.doi.org/10.1080/17588928.2011.604725

Abstract: Although attention may play a major role in explaining EBA/FBA activation during high-order, body-related tasks, it is important to establish the functional significance of top-down modulation in different tasks. While neuroimaging studies documented the functional and anatomical specificity of EBA/FBA activation during body form perception, repetitive transcranial magnetic stimulation (rTMS) and brain-lesion studies provided causative evidence that activity in EBA is essential for processing morphological details of body parts. Local processing of body shapes in EBA might contribute to the representation of high-order body attributes, including person identity and body esthetics, which probably rely on a widespread network of different interconnected areas.

After the first descriptions, 10 years ago, of body-selective activations in the occipitotemporal cortex (Downing, Jiang, Shuman, & Kanwisher, 2001), several studies have now documented modulations of EBA and/or FBA in different tasks involving the visual processing of whole human bodies and non-facial body parts. In their discussion paper, Downing and Peelen review most of this evidence and suggest that the variety of functions involving the body-selective activation of EBA/FBA points to a basic, but still body-specific, perceptual role. Moreover, they propose that the modulation of EBA/FBA neural activity during high-order processing of body stimuli might be attributed to increased attention to the body triggered by task demands. The attentional account is extremely powerful in that it may immediately explain EBA/FBA involvement in high-order functions found in several studies.

We believe, however, that the functional significance of top-down modulation in the different tasks needs to be established. Indeed, attentional top-down modulations of neural activity may be epiphenomenal to task performance, reflecting the increased perceptual saliency of body stimuli in the visual scene or the activity of high-order interconnected areas. Alternatively, it might have a specific functional significance, reflecting an increased recruitment of areas that process stimulus features critical for task performance.

Disentangling whether increased activation of EBA/FBA during high-order, body-related tasks is functionally essential, rather than epiphenomenal, requires direct testing. It is then critical to understand the following: (1) which body-related, high-order tasks modulate activity in EBA/FBA; (2) to what extent different body-related areas in temporal, parietal, and premotor regions are modulated by high-order tasks; (3) which high-order functions are impaired after damage to EBA/FBA.

Previous neuroimaging studies have provided evidence that modulation of EBA/FBA during at least some high-order cognitive functions is functionally (point 1) and anatomically (point 2) specific. For example, while it is likely that a moving body is more interesting than a still body, EBA activity is not modulated by the movement of the observed body (Downing et al., 2001). Conversely, EBA activity is strongly influenced by distortions in body size (Mohr et al., 2011) and is altered in patients with eating disorders (Suchan et al., 2010; Uher et al., 2005). Furthermore, while perception of incoherent, as compared to coherent, sequences of body postures induces greater activation of EBA, the opposite pattern was found in premotor and parietal areas involved in action representation (Downing, Peelen, Wiggett, & Tew, 2006).

An important contribution to understanding the functional significance of EBA/FBA relies on testing the consequences of their lesion in brain-damaged patients or in neurologically intact individuals by repetitive transcranial magnetic stimulation (rTMS) (point 3) (Avenanti & Urgesi, 2011). Dysfunctional activity of EBA/FBA is associated with deficits in the visual discrimination of body parts, but not of face or object parts (Moro et al., 2008; Pitcher, Charles, Devlin, Walsh, & Duchaine, 2009; Urgesi, Berlucchi, & Aglioti, 2004). Importantly, lesions to EBA/FBA led to poor performance when discriminating the morphology of the body parts of different individuals, while performance in discriminating limb postures with different possible or impossible implied actions was not affected (Candidi, Urgesi, Ionta, & Aglioti, 2008; Moro et al., 2008; Urgesi, Candidi, Ionta, & Aglioti, 2007).

This would suggest that EBA may help to discriminate different body identities, probably by processing the local details of body parts which are stable across the range of postural configurations largely shared by all

The authors are supported by grants from Ministero Italiano della Salute (Progetto Giovani Ricercatori GR-2008-1137139; to C.U.) and the Istituto Italiano di Tecnologia (SEED 2009 Protocol Number 21538 to C. U. and A. A.).

human bodies. Accordingly, rTMS of EBA disrupted the local processing of inverted body postures, but not the configural processing of upright bodies (Urgesi, Calvo-Merino, Haggard, & Aglioti, 2007). Local versus configural body processing may be involved in different perceptual and high-order tasks, including person discrimination (Urgesi, Candidi et al., 2007) and evaluation of body esthetics (Calvo-Merino, Urgesi, Orgs, Aglioti, & Haggard, 2010). Claiming that EBA/FBA is involved in a high-order body function does not necessarily imply that it explicitly and specifically represents high-order body attributes (such as identity, action, emotion, or esthetics), but rather that EBA/FBA contributes to the representation of such attributes.

It is likely that high-order body attributes are not unitarily represented in a single brain area, but involve a network of regions, each processing specific aspects. For example, body identity may emerge from the integration of different static (e.g., body shape) and dynamic (e.g., gait) perceptual cues with person-specific "semantic" knowledge, including name, occupation, relationships, etc. All these types of information are processed in different interconnected areas, whose pattern of activation might lead to the representation of person-specific identity. This view is in keeping with Downing and Peelen's proposal, since EBA/FBA is not likely to explicitly represent person identities as whole and unitary attributes (i.e., my romantic partner). This highlights, however, a widespread representation of high-order functions which involves several areas, including EBA/FBA, in processing body shape (the shape of my partner's body).

Understanding which high-order functions involve EBA or FBA activity will advance our knowledge of body visual representations as well as of the multi-faceted organization of complex, body-related functions.

* * *

Faces and bodies in the brain

Giovanni Berlucchi

Neurological Sciences, University of Verona, Verona, Italy

E-mail: giovanni.berlucchi@univr.it

http://dx.doi.org/10.1080/17588928.2011.613986

Abstract: The discovery of face-sensitive and body-sensitive regions in the extrastriate human cortex has raised the problem of the relations of these areas to face and body knowledge and their role in person identification. In this commentary, I point to some as yet unexplored aspects of these cortical regions, including their status as proper anatomo-functional areas, the role of body appearance in the recognition of persons, and the development of body-related and face-related areas in sighted and congenitally blind individuals.

Downing and Peelen (henceforth D&P) present an extensive review of the explosion of studies on body-selective activities of the occipito-temporal cortex which has followed the discovery of the extrastriate body area (EBA) and the fusiform body area (FBA). Both of these areas respond to the visual presentation of bodies and bodily parts except the face. D&P argue that activities in these areas are sufficient to describe the shape and posture of bodies in some detail. More elaborate analyses of bodily structure, actions, and emotional expressions, eventually leading to person identification, are thought by D&P to be carried out in other parts of the brain. This reasonable hypothesis does not exclude the possibility that inputs from EBA and FBA to the putative higher-order areas, as well as feedback from the latter areas to the former, are essential for person identification from bodily appearance and related cognitive abilities. Unfortunately, the psychophysiology of the identification of a person from his/her bodily rather than facial features is not well understood. Bodies are usually covered by clothes and carry paraphernalia which may contribute to or even be essential for person identification, and the face is regarded by most as the paramount if not the only part of human anatomy which allows the recognition of an individual. In her book on visual agnosia, Martha Farah (2004) mentions the old story of the professor walking by two naked students sunbathing, the smarter of whom covers her face instead of her pudenda to avoid being recognized. There is a well-known neurological syndrome of prosopagnosia, the inability to recognize familiar faces, but I am unaware of neurological disorders of person identification specifically based on a faulty appreciation of bodily appearance. In what follows, I will comment on a few questions that have come to my mind while reading the review and will attempt to point to possible experiments that may answer them.

So far, we have little or no cytoarchitectural or hodological information that may confer on EBA and FBA the status of cortical areas in a strictly anatomo-functional sense. As acknowledged by D&P, the work of Orlov, Makin, and Zohary (2010) suggests that EBA and FBA, as identified by fMRI, may simply represent peaks of activity within a broader cortical territory involved in the processing of environmental information, including the category of bodies and bodily parts, but extending beyond it. Since EBA is known to respond not only to visual inputs but also to proprioceptive inputs from the body (e.g., Astafiev, Stanley, Shulman, & Corbetta, 2004), its characterization as a cortical area would be strengthened by the demonstration of selective inputs to it from the somatosensory thalamus or cortex. In vivo studies with diffusion tensor imaging (DTI) have the potential to provide this information.

Although face recognition is a slowly developing ability (Sugita, 2009), it has long been known that newborn infants are able to distinguish a proper face from a "scrambled" face in which the relative positions of eyes, nose, and mouth are rearranged in a random fashion. By contrast, the ability to distinguish proper bodies from scrambled bodies does not emerge until the middle of the second year, suggesting that knowledge of the basic bodily structure is acquired through postnatal neural maturation or experience or both (Slaughter & Heron, 2004). In affective social exchanges with their mothers, babies most probably pay more attention to emotional visual and acoustic signals from the mother's face than to signals from her body. Further, the capacity of newborn babies to imitate gestures (Meltzoff & Moore, 1977), if it exists, is limited to the orofacial territory and possibly only to tongue protrusion (Jones, 2009). All this evidence suggests a possibly differential development of face-sensitive and body-sensitive regions in the human cortex, and I wonder if there is any electrophysiological or fMRI evidence to prove it.

Konorski (1967) was the first to postulate the existence of separate "gnostic units" (neurons or cortical regions) specifically reactive to various categories of visual stimuli, one of which is the seen position of one's own limbs. He thought that the putative gnostic units for the latter category are utilized very early in life to build up associations between visual inputs from visible body parts and the proprioceptive sensations from those same parts. Such a combination of visual and somatic inputs may be necessary to construct a correct internal image of one's own body as well as of other bodies. Congenitally blind children draw and model human bodies less accurately than blindfolded sighted children with regard to both presence and placement of body parts, as well as an exaggerated size attributed to hands and arms, which are their main means of space exploration (Critchley, 1953; Kinsbourne & Lempert, 1980). In sighted adults, processing of the body by EBA and FBA extends beyond the visual modality, since both areas are activated by tactile exploration of human faces and other body parts such as hands and feet (Kitada, Johnsrude, Kochiyama, & Lederman, 2009). Further, again in sighted adults, EBA is activated by one's own movements made out of view (Astafiev et al., 2004), as well as by haptic and visual imagery of body parts. (Costantini, Urgesi, Galati, Romani, & Aglioti, 2011). It is known that in congenitally blind adults different forms of space perception and representation engage posterior brain areas endowed with visual functions in sighted individuals (e.g., Renier et al., 2010; Ricciardi et al., 2009). I wonder whether face- and body-sensitive areas, including EBA and FBA, can be localized with such stimuli and procedures in congenitally blind adults. If so, one would have to assume a commitment, perhaps even an early one, of these areas to represent the body independently of vision.

* * *

 http://dx.doi.org/10.1080/17588928.2011.613987

Reply to Commentaries

How might occipitotemporal body-selective regions interact with other brain areas to support person perception?

Paul E. Downing[1] and Marius V. Peelen[2]

[1]Wales Institute of Cognitive Neuroscience, School of Psychology, Bangor University, Bangor, UK
[2]Center for Mind/Brain Sciences, University of Trento, Rovereto, Italy

We argued in our Discussion Paper for the view that the main functional role of occipitotemporal body-selective regions is to make explicit, via patterns of activity, the shape and posture of perceived bodies, rather than directly representing higher-order person attributes. Much of the commentary was on the question of how activity in other brain areas interacts with activity in EBA and FBA. In this reply, we emphasize that our claims do not imply that EBA and FBA are cognitively impenetrable modules that are driven only by bottom-up input from earlier visual regions. Instead, it is likely that these regions interact heavily with other brain regions, and that their activity is shaped—in limited ways—by both feed-forward and feedback connections. In the context of such large-scale networks, EBA and FBA activity will be most effective at conveying detailed information about the shape and posture of bodies in the current percept.

One common theme was on the minds of the majority of commentators: How do EBA and FBA interact with other areas of the brain, and, in particular, how does the activity of higher-order areas influence and shape the representations in these body-selective regions? Various commentators approach this in different ways and with different emphases. For example, Berlucchi states that "inputs from EBA and FBA to the putative higher-order areas, as well as feedbacks from the latter areas to the former, are essential for person identification from bodily appearance and related cognitive abilities"; Ewbank states that "the relationship between EBA and FBA, and the relationship between these regions and higher-level areas, is likely to be reciprocal, with activity in higher-levels areas dynamically shaping responsivity in lower-levels regions"; and Ramsey et al. "propose that [EBA and FBA] are a part of a reciprocally connected cortical network that functions to minimize prediction error when making identity inferences."

Several commentators (Berlucchi; Quadflieg & Rossion; Urgesi & Avenanti) explicitly note that an interaction between higher-order areas and EBA/FBA is compatible with our main proposal that these regions encode the shapes and postures of bodies. However, such interactions were also raised as a challenge to our proposal that representations in EBA/FBA are "cognitively unelaborated" (see commentaries by Ewbank; de Lange & Bekkering; Quadflieg & Rossion; and Ramsey, Cross, & van Schie). In our Discussion Paper, we did not discuss putative interactions between EBA/FBA and other brain regions in detail, mostly because there is not much evidence to go by (other than evidence for attentional modulation). We speculated (p. 193) that activity in EBA and FBA likely follows the subjective percept rather than the retinal image, and that these regions may therefore be influenced by person identity inferences made in other regions (cf. commentaries of Quadflieg & Rossion and Ramsey et al.). Furthermore, we noted (pp. 197, 199) that visual imagery (and presumably working memory) may activate body representations in EBA/FBA (cf. commentary of de Lange & Bekkering). Finally, rather than proposing that EBA and FBA are "cognitively impenetrable modules" (Ewbank), we proposed (p. 202) that there are good functional arguments for why representations in the perceptual nodes of a network should not be fully penetrable by other parts of the network, again stressing the lack of empirical work in this area. As we will outline below, we support the idea that higher-order regions influence representations in EBA and FBA beyond attentional modulation (e.g., as in the case of visual imagery). We argue that such top-down influences are nonetheless compatible with our view that these regions primarily encode body shape and body posture rather than higher-order person attributes.

It may be useful to think of EBA and FBA as forming an ad hoc network with other remote brain regions when the current goals or focus of interest is relevant to body perception. The "nodes" of this network each contribute information to the whole, in the form of its patterns of activity. The activity of these nodes will bias and influence each other, so that the network will normally tend to settle on a coherent, mutually consistent state that results in some kind of understanding of the scene. In this context, EBA and FBA activity will, as we have argued, be most effective at conveying detailed information about the shape and posture of the bodies of people in the scene.

While in some ways this representation of body shape and posture might be considered a "low-level" representation, we do not argue that EBA and FBA produce a mere copy of the aspects of the retinal image that come from bodies. Instead, we expect that patterns of activity relate to the subjective body percept, which arises from coordinated activity of many areas. Here are two concrete examples (to our knowledge, the relevant studies do not exist). First, by analogy to "Mooney" (1957) faces, we would expect responses to bodies/body parts to increase in FBA and EBA when a highly schematized image produces a body percept, compared to when it does not. Experiencing the percept of a body will shape the way that individual reduced or ambiguous features are interpreted with a corresponding effect on patterns of activity in these regions. Second, prior knowledge about a perceived individual can strongly influence the percept of a person, and likewise the patterns of activity in EBA/FBA. For example, consider a scenario where a person is seen but is partially occluded or otherwise covered, as by loose clothing. Where that person is known to the observer, we would expect a more detailed, focused pattern of response in EBA/FBA reflecting the knowledge of that shape, compared to an unknowing observer.

This second example in particular raises the question of how, and to what extent, body representations in EBA and FBA are "penetrable"; that is, how their activity is modulated or shaped by the influences of other brain regions. The simplest possibility is that overall activity in these regions is grossly up- or down-regulated via such mechanisms as spatial attention (e.g., attending toward or away from the person in a scene). In our Discussion Paper, we argued for such a mechanism as an alternative account to explain modulations that authors had taken as evidence for more complex processes in these regions. More subtly, similar mechanisms could come into play to select among possible distinct body representations. For example, switching attention from someone's hand to their foot may not change overall activity levels in EBA but may influence the detailed pattern of activity that is expressed in that region. Finally, activity in other areas may help to "complete" a missing percept, or, similarly, to form a prediction about an upcoming percept. To return to our earlier example: Where an unknown person in an unfamiliar setting is seen, but is partly occluded, local associative processes may be sufficient to complete a general but imprecise representation that would "fill in" the missing stimulus details. Where contextual information or other cues about the identity of the person are available (e.g., through the face), these may further support such completion processes, allowing a more detailed and accurate completion of the missing information.

These proposals allow EBA and FBA to play a rich and complex role in the perception of other people, in combination with other areas, yet preserve the parsimonious idea that the patterns of activity local to these areas are tuned to the relatively simple properties of shape and posture. They show how these patterns can work interactively to extract, for example, the identity of the people around us, without directly and locally coding for identity *per se*. Of course, these questions of connectivity and penetrability are not specific to body representation. In neuroimaging studies, many of the recent efforts to answer these questions have focused, for some good reasons, on faces (e.g., Fairhall & Ishai, 2007). As more becomes known about the neural encoding of the appearance of the body, EBA and FBA will provide a valuable additional test bed for hypotheses about local encoding and long-range influences (Ewbank et al., 2011).

Aside from considering connectivity and penetrability of representations, several of the commentaries explore other more specific issues. We focus briefly on two: the distinction between body parts and wholes, and the development of body representations.

Chan and Baker argue that the evidence for a part-based bias to body representations in EBA, relative to FBA, is not iron-clad. We agree, although we note other TMS evidence that is also consistent with this idea (e.g., Urgesi, Calvo-Merino, Haggard, & Aglioti, 2007). More broadly, there is increasing evidence that distinctions between the functional properties of EBA and FBA may be usefully understood in reference to similar focal extrastriate representations of faces and of object form more generally. That is, for each of these kinds, fMRI studies show focal, selective activations that are relatively posterior-lateral (EBA, face-selective OFA, and object-form selective LO) and distinct foci that are relatively ventro-medial (FBA, face-selective FFA, and object-form selective pFs). Across categories, these representations appear to be distinguished in several ways, including among them a bias for part over whole representations in the posterior-lateral regions, and the reverse bias in ventro-medial representations (Taylor & Downing, 2011).

Further, in an earlier paper, Taylor, Wiggett, and Downing (2007) noted that the hierarchical organization of the body into parts is systematic ("fingers," "arms," etc.) and asked whether EBA body representations might respect these boundaries, carving at the joints, as it were. Kemmerer valuably points out that linguistic partitions of the body are not culturally universal, so that, at some level at least, what is a natural "part" of the body for one individual may not be so for another. This provokes an interesting question about the susceptibility of part representations in EBA to semantic/linguistic influences. In the spirit of our proposed framework, we might expect no such influence. However, it is possible that these semantic /linguistic influences shape the way that individuals attend to the bodies of others in daily life. If these biases are regular and consistent, it is conceivable that they would in turn shape, over time, the "space" of neural patterns representing the body in these regions.

Slaughter and Berlucchi both comment on the development of body structure knowledge in infants. Evidence for late development of body structure knowledge comes from studies measuring the time infants look at intact versus scrambled bodies (e.g., a body with arms attached to the hips). Initial studies estimated that it is not before 15 months of age that infants' looking behavior discriminates between intact and scrambled bodies (Slaughter & Heron, 2004). In a later study (Heron & Slaughter, 2010), this was shown to depend on the realism of the stimuli, with the earliest age of discriminating realistic intact from realistic scrambled bodies being around 9 months of age. By contrast, sensitivity to body motion and goal-directed action appears to develop earlier in life (Bertenthal, 1993; Király, Jovanovic, Prinz, Aschersleben, & Gergely, 2003). According to Slaughter, this developmental dissociation is consistent with our proposed dissociation between EBA and FBA (involved in the encoding of body shape) and other regions involved in the encoding of body motion and action goals. She further suggests that the developmental dissociation implies that certain aspects of higher-level person perception can occur without the information computed by EBA and FBA. While this is an interesting idea, what needs to be shown to complete the link that Slaughter makes is that developing knowledge of body structure relates systematically to increasing sophistication of the body shape/posture representations in EBA/FBA. Without this, it is still unclear whether the relatively late development of body structure knowledge in infants should be interpreted as providing evidence that high-level person perception, such as perceiving action goals, can bypass EBA and FBA. Finally, it is not unlikely that future studies, using more sensitive methods, may reveal body structure knowledge in infants at increasingly younger ages.

To sum up, the simple model of body encoding that we proposed in our Discussion Paper explains a great deal of the current literature and does so parsimoniously. When considering how EBA and FBA work interactively with other brain areas, as they must, we can see how their shape and posture representations could contribute critically to many high-level functions of person perception, such as the identification of others. Much remains to be learned about how these two areas make distinct contributions to person perception. More generally, it is clear that in the next few years there will be a lot of energy directed at the broad question of how the nodes of the brain's interconnected networks influence each other. The highly focal, well-characterized nature of extrastriate body representations, and the many cognitive and perceptual functions for which the perception of other people is relevant, make these areas worthy of a focus in these ongoing studies.

References from the Discussion Paper, the Commentaries, and the Reply

Aguirre, G. K., Zarahn, E., & D'Esposito, M. (1998). An area within human ventral cortex sensitive to "building" stimuli: Evidence and implications. *Neuron*, *21*(2), 373–383.

Aleong, R., & Paus, T. (2010). Neural correlates of human body perception. *Journal of Cognitive Neuroscience*, *22*(3), 482–495.

Allison, T., Puce, A., & McCarthy, G. (2000). Social perception from visual cues: Role of the STS region. *Trends in Cognitive Sciences*, *4*(7), 267–278.

Amaral, D. G., Behniea, H., & Kelly, J. L. (2003). Topographic organization of projections from the amygdala to the visual cortex in the macaque monkey. *Neuroscience*, *118*(4), 1099–1120.

Amedi, A., von Kriegstein, K., Van Atteveldt, N. M., Beauchamp, M. S., & Naumer, M. J. (2005). Functional imaging of human crossmodal identification and object recognition. *Experimental Brain Research*, *166*(3–4), 559–571.

Anderson, A. K. (2005). Affective influences on the attentional dynamics supporting awareness. *Journal of Experimental Psychology: General*, *134*(2), 258–281.

Appelbaum, L. G., Wade, A. R., Vildavski, V. Y., Pettet, M. W., & Norcia, A. M. (2006). Cue-invariant networks for figure and background processing in human visual cortex. *Journal of Neuroscience*, *26*, 11695–11708.

Arzy, S., Thut, G., Mohr, C., Michel, C. M., & Blanke, O. (2006). Neural basis of embodiment: Distinct contributions of temporoparietal junction and extrastriate body area. *Journal of Neuroscience*, *26*(31), 8074–8081.

Aslin, R. (2009). How infants view natural scenes gathered from a head-mounted camera. *Optometry and Vision Science*, *86*, 561–565.

Astafiev, S. V., Stanley, C. M., Shulman, G. L., & Corbetta, M. (2004). Extrastriate body area in human occipital cortex responds to the performance of motor actions. *Nature Neuroscience*, *7*(5), 542–548.

Astafiev, S. V., Stanley, C. M., Shulman, G. L., & Corbetta, M. (2005). Is the extrastriate body area involved in motor actions? Reply. *Nature Neuroscience*, *8*(2), 125–126.

Atkinson, A. P., Dittrich, W. H., Gemmell, A. J., & Young, A. W. (2004). Emotion perception from dynamic and static body expressions in point-light and full-light displays. *Perception*, *33*(6), 717–746.

Avenanti, A., & Urgesi, C. (2011). Understanding 'what' others do: Mirror mechanisms play a crucial role in action perception. *Social Cognitive and Affective Neuroscience*, *6*, 257–259.

Avikainen, S., Liuhanen, S., Schurmann, M., & Hari, R. (2003). Enhanced extrastriate activation during observation of distorted finger postures. *Journal of Cognitive Neuroscience*, *15*(5), 658–663.

Bach, P., Peelen, M. V., & Tipper, S. P. (2010). On the role of object information in action observation: An fMRI study. *Cerebral Cortex*, *20*(12), 2798–2809.

Bar, M. (2009). The proactive brain: Memory for predictions. *Philosophical Transactions of the Royal Society of London, Series B: Biological Sciences*, *364*, 1235–1243.

Barsalou, L.W. (2008). Grounded cognition. *Annual Review of Psychology*, *59*, 617–645.

Beauchamp, M. S., Lee, K. E., Haxby, J. V., & Martin, A. (2002). Parallel visual motion processing streams for manipulable objects and human movements. *Neuron*, *34*(1), 149–159.

Bedny, M., Caramazza, A., Grossman, E., Pascual-Leone, A., & Saxe, R. (2008). Concepts are more than percepts: The case of action verbs. *Journal of Neuroscience*, *28*(44), 11347–11353.

Bekkering, H., & Neggers, S. F. (2002). Visual search is modulated by action intentions. *Psychological Science*, *13*(4), 370–374.

Bell, A. J., & Sejnowski, T. J. (1997). The "independent components" of natural scenes are edge filters. *Vision Research*, *377*, 3327–3338.

Benuzzi, F., Lui, F., Duzzi, D., Nichelli, P. F., & Porro, C. A. (2008). Does it look painful or disgusting? Ask your parietal and cingulate cortex. *Journal of Neuroscience*, *28*(4), 923–931.

Berlucchi, G., & Aglioti, S. M. (2010). The body in the brain revisited. *Experimental Brain Research*, *200*(1), 25–35.

Bertenthal, B. (1993). Infants' perception of biomechanical motions: Intrinsic image and knowledge-based constraints. In C. E. Granrud (Ed.), *Visual perception and cognition in infancy* (pp. 175–214). Hillsdale, NJ: Lawrence Erlbaum Associates.

Blake, R., & Shiffrar, M. (2007). Perception of human motion. *Annual Review of Psychology*, *58*, 47–73.

Blanke, O., Ionta, S., Fornari, E., Mohr, C., & Maeder, P. (2010). Mental imagery for full and upper human bodies: Common right hemisphere activations and distinct extrastriate activations. *Brain Topography*, *23*(3), 321–332.

Bracci, S., Ietswaart, M., Peelen, M. V., & Cavina-Pratesi, C. (2010). Dissociable neural responses to hands and non-hand body parts in human left extrastriate visual cortex. *Journal of Neurophysiology*, *103*(6), 3389–3397.

Brandman, T., & Yovel, G. (2010). The body inversion effect is mediated by face-selective, not body-selective, mechanisms. *Journal of Neuroscience*, *30*(31), 10534–10540.

Brefczynski, J. A., & DeYoe, E. A. (1999). A physiological correlate of the 'spotlight' of visual attention. *Nature Neuroscience*, *2*(4), 370–374.

Bruce, V., & Young, A. (1986). Understanding face recognition. *British Journal of Psychology*, *77*(Pt 3), 305–327.

Bruner, J. (1973). *Going beyond the information given*. New York, NY: Norton.

Calvo-Merino, B., Urgesi, C., Orgs, G., Aglioti, S. M., & Haggard, P. (2010). Extrastriate body area underlies aesthetic evaluation of body stimuli. *Experimental Brain Research*, *204*(3), 447–456.

Candidi, M., Urgesi, C., Ionta, S., & Aglioti, S. M. (2008). Virtual lesion of ventral premotor cortex impairs visual perception of biomechanically possible but not impossible actions. *Social Neuroscience*, *3*(3–4), 388–400.

Carter, E. J., & Pelphrey, K. A. (2008). Friend or foe? Brain systems involved in the perception of dynamic signals of menacing and friendly social approaches. *Social Neuroscience*, *3*(2), 151–163.

Chaminade, T., Meltzoff, A. N., & Decety, J. (2005). An fMRI study of imitation: Action representation and body schema. *Neuropsychologia, 43*(1), 115–127.

Chan, A. W., Kravitz, D. J., Truong, S., Arizpe, J., & Baker, C. I. (2010). Cortical representations of bodies and faces are strongest in commonly experienced configurations. *Nature Neuroscience, 13*(4), 417–418.

Chan, A. W., Peelen, M. V., & Downing, P. E. (2004). The effect of viewpoint on body representation in the extrastriate body area. *Neuroreport, 15*(15), 2407–2410.

Chao, L. L., Haxby, J. V., & Martin, A. (1999). Attribute-based neural substrates in temporal cortex for perceiving and knowing about objects. *Nature Neuroscience, 2*, 913–919.

Cikara, M., Eberhardt, J. L., & Fiske, S. T. (2011). From agents to objects: Sexist attitudes and neural responses to sexualized targets. *Journal of Cognitive Neuroscience, 23*(3), 540–551.

Costantini, M., Galati, G., Ferretti, A., Caulo, M., Tartaro, A., Romani, G. L., et al. (2005). Neural systems underlying observation of humanly impossible movements: An fMRI study. *Cerebral Cortex, 15*(11), 1761–1767.

Costantini, M., Urgesi, C., Galati, G., Romani, G. L., & Aglioti, S. M. (2011). Haptic perception and body representation in lateral and medial occipito-temporal cortices. *Neuropsychologia, 49*(5), 821–829.

Critchley, M. (1953). Tactile thought, with special reference to the blind. *Brain, 76*, 19–35.

Cross, E. S., Mackie, E. C., Wolford, G., & Hamilton, A. F. (2010). Contorted and ordinary body postures in the human brain. *Experimental Brain Research, 204*(3), 397–407.

Danckert, J. A., & Goodale, M. A. (2003). Ups and downs in the visual control of action. In S. H. Johnson-Frey (Ed.), *Taking action: Cognitive neuroscience perspectives on intentional acts* (pp. 29–64). Cambridge, MA: MIT Press.

David, N., Cohen, M. X., Newen, A., Bewernick, B. H., Shah, N. J., Fink, G. R., et al. (2007). The extrastriate cortex distinguishes between the consequences of one's own and others' behavior. *NeuroImage, 36*(3), 1004–1014.

David, N., Jansen, M., Cohen, M. X., Osswald, K., Molnar-Szakacs, I., Newen, A., et al. (2009). Disturbances of self–other distinction after stimulation of the extrastriate body area in the human brain. *Social Neuroscience, 4*(1), 40–48.

David, S. V., Hayden, B. Y., Mazer, J. A., & Gallant, J. L. (2008). Attention to stimulus features shifts spectral tuning of V4 neurons during natural vision. *Neuron, 59*, 509–521.

De Lange, F. P., Spronk, M., Willems, R. M., Toni, I., & Bekkering, H. (2008). Complementary systems for understanding action intentions. *Current Biology, 18*(6), 454–457.

Desimone, R., Albright, T. D., Gross, C. G., & Bruce, C. (1984). Stimulus-selective properties of inferior temporal neurons in the macaque. *Journal of Neuroscience, 4*(8), 2051–2062.

Devue, C., Collette, F., Balteau, E., Dequeldre, C., Luxen, A., Maquet, P., et al. (2007). Here I am: The cortical correlates of visual self-recognition. *Brain Research, 1143*, 169–182.

Dijkerman, H. C., McIntosh, R. D., Schindler, I., Nijboer, T. C. W., & Milner, A. D. (2009). Choosing between alternative wrist postures: Action planning needs perception. *Neuropsychologia, 477*, 1476–1482.

Dow, B. M., Snyder, A. Z., Vautin, R. G., & Bauer, R. (1981). Magnification factor and receptive field size in foveal striate cortex of the monkey. *Experimental Brain Research, 447*, 213–228.

Downing, P., Liu, J., & Kanwisher, N. (2001). Testing cognitive models of visual attention with fMRI and MEG. *Neuropsychologia, 39*(12), 1329–1342.

Downing, P. E., Chan, A. W., Peelen, M. V., Dodds, C. M., & Kanwisher, N. (2006). Domain specificity in visual cortex. *Cerebral Cortex, 16*(10), 1453–1461.

Downing, P. E., Jiang, Y., Shuman, M., & Kanwisher, N. G. (2001). A cortical area selective for visual processing of the human body. *Science, 293*(5539), 2470–2473.

Downing, P. E., Peelen, M. V., Wiggett, A. J., & Tew, B. D. (2006). The role of the extrastriate body area in action perception. *Social Neuroscience, 1*(1), 52–62.

Downing, P. E., Wiggett, A. J., & Peelen, M. V. (2007). Functional magnetic resonance imaging investigation of overlapping lateral occipitotemporal activations using multi-voxel pattern analysis. *Journal of Neuroscience, 27*(1), 226–233.

Drucker, D. M., & Aguirre, G. K. (2009). Different spatial scales of shape similarity representation in lateral and ventral LOC. *Cerebral Cortex, 19*(10), 2269–2280.

Epstein, R., & Kanwisher, N. (1998). A cortical representation of the local visual environment. *Nature, 392*(6676), 598–601.

Ewbank, M. P., Lawson, R. P., Henson, R. N., Rowe, J. B., Passamonti, L., & Calder, A. J. (2011). Changes in "top-down" connectivity underlie repetition suppression in the ventral visual pathway. *Journal of Neuroscience, 31*(15), 5635–5642.

Fairhall, S. L., & Ishai, A. (2007). Effective connectivity within the distributed cortical network for face perception. *Cerebral Cortex, 17*(10), 2400–2406.

Farah, M. J. (2004). *Visual agnosia*. Cambridge, MA: MIT Press.

Flaisch, T., Schupp, H. T., Renner, B., & Junghofer, M. (2009). Neural systems of visual attention responding to emotional gestures. *NeuroImage, 45*(4), 1339–1346.

Friston, K. J. (2005). A theory of cortical responses. *Philosophical Transactions of the Royal Society of London, Series B: Biological Sciences, 360*(1456), 815–836.

Gandhi, S. P., Heeger, D. J., & Boynton, G. M. (1999). Spatial attention affects brain activity in human primary visual cortex. *Proceedings of the National Academy of Sciences of the United States of America, 96*(6), 3314–3319.

Gauthier, I., Tarr, M. J., Moylan, J., Skudlarski, P., Gore, J. C., & Anderson, A. W. (2000). The fusiform "face area" is part of a network that processes faces at the individual level. *Journal of Cognitive Neuroscience, 12*(3), 495–504.

Georgieff, N., & Jeannerod, M. (1998). Beyond consciousness of external reality: A "who" system for consciousness of action and self-consciousness. *Consciousness and Cognition, 7*(3), 465–477.

Ghuman, A. S., McDaniel, J. R., & Martin, A. (2010). Face adaptation without a face. *Current Biology, 20*(1), 32–36.

Giese, M. A., & Poggio, T. (2003). Neural mechanisms for the recognition of biological movements. *Nature Reviews Neuroscience*, *4*(3), 179–192.

Grafton, S. T., & Hamilton, A. F. (2007). Evidence for a distributed hierarchy of action representation in the brain. *Human Movement Science*, *26*(4), 590–616.

Graziano, M. S., & Aflalo, T. N. (2007). Rethinking cortical organization: Moving away from discrete areas arranged in hierarchies. *Neuroscientist*, *13*(2), 138–147.

Grezes, J., Pichon, S., & de Gelder, B. (2007). Perceiving fear in dynamic body expressions. *NeuroImage*, *35*(2), 959–967.

Grill-Spector, K., Henson, R., & Martin, A. (2006). Repetition and the brain: Neural models of stimulus-specific effects. *Trends in Cognitive Science*, *10*(1), 14–23.

Grill-Spector, K., Kushnir, T., Edelman, S., Avidan, G., Itzchak, Y., & Malach, R. (1999). Differential processing of objects under various viewing conditions in the human lateral occipital complex. *Neuron*, *24*(1), 187–203.

Grosbras, M. H., & Paus, T. (2006). Brain networks involved in viewing angry hands or faces. *Cerebral Cortex*, *16*(8), 1087–1096.

Gross, C. G. (1992). Representation of visual stimuli in inferior temporal cortex. *Philosophical Transactions of the Royal Society of London. Series B, Biological Sciences*, *335*(1273), 3–10.

Gross, C. G., Bender, D. B., & Rocha-Miranda, C. E. (1969). Visual receptive fields of neurons in inferotemporal cortex of the monkey. *Science*, *166*(910), 1303–1306.

Gross, C. G., Rocha-Miranda, C. E., & Bender, D. B. (1972). Visual properties of neurons in inferotemporal cortex of the macaque. *Journal of Neurophysiology*, *35*(1), 96–111.

Grossman, E. D., & Blake, R. (2002). Brain areas active during visual perception of biological motion. *Neuron*, *35*(6), 1167–1175.

Gu, X., Liu, X., Guise, K. G., Naidich, T. P., Hof, P. R., & Fan, J. (2010). Functional dissociation of the frontoinsular and anterior cingulate cortices in empathy for pain. *Journal of Neuroscience*, *30*(10), 3739–3744.

Hadjikhani, N., & de Gelder, B. (2003). Seeing fearful body expressions activates the fusiform cortex and amygdala. *Current Biology*, *13*(24), 2201–2205.

Hamann, S., Herman, R. A., Nolan, C. L., & Wallen, K. (2004). Men and women differ in amygdala response to visual sexual stimuli. *Nature Neuroscience*, *7*(4), 411–416.

Harrison, S. A., & Tong, F. (2009). Decoding reveals the contents of visual working memory in early visual areas. *Nature*, *458*(7238), 632–635.

Haxby, J. V., Gobbini, M. I., Furey, M. L., Ishai, A., Schouten, J. L., & Pietrini, P. (2001). Distributed and overlapping representations of faces and objects in ventral temporal cortex. *Science*, *293*(5539), 2425–2430.

Haxby, J. V., Hoffman, E. A., & Gobbini, M. A. (2000). The distributed human neural system for face perception. *Trends in Cognitive Sciences*, *4*, 223–233.

Haynes, J. D., & Rees, G. (2006). Decoding mental states from brain activity in humans. *Nature Reviews Neuroscience*, *7*(7), 523–534.

Henson, R. N. (2003). Neuroimaging studies of priming. *Progress in Neurobiology*, *70*(1), 53–81.

Heron, M., & Slaughter, V. (2010). Infants' responses to real humans and representations of humans. *International Journal of Behavioral Development*, *34*(1), 34–45.

Hinton, G. E., McClelland, J. L., & Rumelhart, D. E. (1986). Distributed representations. In D. E. Rumelhart & J. L. McClelland (Eds.), *Parallel distributed processing: Explorations in the microstructure of cognition* (pp. 77–109). Cambridge, MA: MIT Press.

Hirai, M., & Hiraki, K. (2005). An event-related potentials study of biological motion perception in human infants. *Brain Research. Cognitive Brain Research*, *22*, 301–304.

Hodzic, A., Kaas, A., Muckli, L., Stirn, A., & Singer, W. (2009). Distinct cortical networks for the detection and identification of human body. *NeuroImage*, *45*(4), 1264–1271.

Hodzic, A., Muckli, L., Singer, W., & Stirn, A. (2009). Cortical responses to self and others. *Human Brain Mapping*, *30*(3), 951–962.

Hubel, D. H., & Wiesel, T. N. (1968). Receptive fields and functional architecture of monkey striate cortex. *Journal of Physiology*, *1957*, 215–243.

Ishizu, T., Amemiya, K., Yumoto, M., & Kojima, S. (2010). Magnetoencephalographic study of the neural responses in body perception. *Neuroscience Letters*, *481*(1), 36–40.

Ishizu, T., Noguchi, A., Ito, Y., Ayabe, T., & Kojima, S. (2009). Motor activity and imagery modulate the body-selective region in the occipital-temporal area: A near-infrared spectroscopy study. *Neuroscience Letters*, *465*(1), 85–89.

Jackson, P. L., Meltzoff, A. N., & Decety, J. (2006). Neural circuits involved in imitation and perspective-taking. *NeuroImage*, *31*(1), 429–439.

Jastorff, J., & Orban, G. A. (2009). Human functional magnetic resonance imaging reveals separation and integration of shape and motion cues in biological motion processing. *Journal of Neuroscience*, *29*(22), 7315–7329.

Johansson, G. (1973). Visual perception of biological motion and a model for its analysis. *Perception & Psychophysics*, *14*, 201–211.

Johnson-Frey, S. H., Maloof, F. R., Newman-Norlund, R., Farrer, C., Inati, S., & Grafton, S. T. (2003). Actions or hand–object interactions? Human inferior frontal cortex and action observation. *Neuron*, *39*(6), 1053–1058.

Jones, S. S. (2009). The development of imitation in infancy. *Philosophical Transactions of the Royal Society of London, Series B: Biological Sciences*, *364*, 2325–2335.

Kable, J. W., & Chatterjee, A. (2006). Specificity of action representations in the lateral occipitotemporal cortex. *Journal of Cognitive Neuroscience*, *18*(9), 1498–1517.

Kamitani, Y., & Tong, F. (2005). Decoding the visual and subjective contents of the human brain. *Nature Neuroscience*, *8*(5), 679–685.

Kanwisher, N., McDermott, J., & Chun, M. M. (1997). The fusiform face area: A module in human extrastriate cortex specialized for face perception. *Journal of Neuroscience*, *17*(11), 4302–4311.

Kastner, S., De Weerd, P., Desimone, R., & Ungerleider, L. G. (1998). Mechanisms of directed attention in the human extrastriate cortex as revealed by functional MRI. *Science*, *282*(5386), 108–111.

Kelly, M. H., & Freyd, J. J. (1987). Explorations of representational momentum. *Cognitive Psychology, 19*(3), 369–401.

Kemmerer, D., & Tranel, D. (2008). Searching for the elusive neural substrates of body part terms: A neuropsychological study. *Cognitive Neuropsychology, 25*(4), 601–629.

Kiefer, M., & Pulvermüller, F. (2011). Conceptual representations in mind and brain: Theoretical developments, current evidence, and future directions. *Cortex*. Advance online publication. doi: 10.1016/j.cortex.2011.04.006.

Kilner, J., Friston, K., & Frith, C. (2007). Predictive coding: An account of the mirror neuron system. *Cognitive Processing, 8*(3), 159–166.

Kinsbourne, M., & Lempert, H. (1980). Human figure representation by blind children. *Journal of General Psychology, 102*, 33–37.

Király, I., Jovanovic, B., Prinz, W., Aschersleben, G., & Gergely, G. (2003). The early origins of goal attribution in infancy. *Consciousness and Cognition, 12*, 752–769.

Kitada, R., Johnsrude, I. S., Kochiyama, T., & Lederman, S. J. (2009). Functional specialization and convergence in the occipito-temporal cortex supporting haptic and visual identification of human faces and body parts: An fMRI study. *Journal of Cognitive Neuroscience, 21*(10), 2027–2045.

Kokal, I., Gazzola, V., & Keysers, C. (2009). Acting together in and beyond the mirror neuron system. *NeuroImage, 47*(4), 2046–2056.

Konorski, J. (1967). *Integrative activity of the brain: An interdisciplinary approach*. Chicago, IL: University of Chicago Press.

Kontaris, I., Wiggett, A. J., & Downing, P. E. (2009). Dissociation of extrastriate body and biological-motion selective areas by manipulation of visual-motor congruency. *Neuropsychologia, 47*(14), 3118–3124.

Kourtzi, Z., & Kanwisher, N. (2000). Activation in human MT/MST by static images with implied motion. *Journal of Cognitive Neuroscience, 12*(1), 48–55.

Kravitz, D. J., Kriegeskorte, N., & Baker, C. I. (2010). High-level visual object representations are constrained by position. *Cerebral Cortex, 20*(12), 2916–2925.

Kret, M. E., Pichon, S., Grezes, J., & de Gelder, B. (2011). Similarities and differences in perceiving threat from dynamic faces and bodies. An fMRI study. *NeuroImage, 54*(2), 1755–1762.

Kriegeskorte, N., Formisano, E., Sorger, B., & Goebel, R. (2007). Individual faces elicit distinct response patterns in human anterior temporal cortex. *Proceedings of the National Academy of Sciences of the United States of America, 104*(51), 20600–20605.

Kuhn, S., Keizer, A., Rombouts, S. A., & Hommel, B. (2011). The functional and neural mechanism of action preparation: Roles of EBA and FFA in voluntary action control. *Journal of Cognitive Neuroscience, 23*(1), 214–220.

Lamm, C., & Decety, J. (2008). Is the extrastriate body area (EBA) sensitive to the perception of pain in others? *Cerebral Cortex, 18*(10), 2369–2373.

Lamme, V. A. (2006). Towards a true neural stance on consciousness. *Trends in Cognitive Sciences, 107*, 494–501.

Lamme, V. A., & Roelfsema, P. R. (2000). The distinct modes of vision offered by feedforward and recurrent processing. I. *Trends in Neuroscience, 237*, 571–579.

Lange, J., & Lappe, M. (2006). A model of biological motion perception from configural form cues. *Journal of Neuroscience, 26*(11), 2894–2906.

Lee, T. S., Mumford, D., Romero, R., & Lamme, V. A. (1998). The role of the primary visual cortex in higher level vision. *Vision Research, 387*, 2429–2454.

Lee, T. S., & Nguyen, M. (2001). Dynamics of subjective contour formation in the early visual cortex. *Proceedings of the National Academy of Sciences of the United States of America, 987*, 1907–1911.

Leube, D. T., Knoblich, G., Erb, M., Grodd, W., Bartels, M., & Kircher, T. T. (2003). The neural correlates of perceiving one's own movements. *NeuroImage, 20*(4), 2084–2090.

Levy, I., Hasson, U., Avidan, G., Hendler, T., & Malach, R. (2001). Center-periphery organization of human object areas. *Nature Neuroscience, 4*(5), 533–539.

Liu, J., Harris, A., & Kanwisher, N. (2010). Perception of face parts and face configurations: An fMRI study. *Journal of Cognitive Neuroscience, 22*(1), 203–211.

Lorteije, J. A., Barraclough, N. E., Jellema, T., Raemaekers, M., Duijnhouwer, J., Xiao, D., et al. (2011). Implied motion activation in cortical area MT can be explained by visual low-level features. *Journal of Cognitive Neuroscience, 23*(6), 533–1548.

Macrae, C. N., & Quadflieg, S. (2010). Perceiving people. In S. Fiske, D. T. Gilbert, & G. Lindzey (Eds.), *The handbook of social psychology* (5th ed., pp. 428–463). New York, NY: McGraw-Hill.

Mahon, B. Z., Anzellotti, S., Schwarzbach, J., Zampini, M., & Caramazza, A. (2009). Category-specific organization in the human brain does not require visual experience. *Neuron, 63*(3), 397–405.

Majid, A. (2010). Words for parts of the body. In B. Malt & P. Wolff (Eds.), *Words and the mind* (pp. 58–71). Oxford, UK: Oxford University Press.

Malach, R., Reppas, J. B., Benson, R. R., Kwong, K. K., Jiang, H., Kennedy, W. A., et al. (1995). Object-related activity revealed by functional magnetic resonance imaging in human occipital cortex. *Proceedings of the National Academy of Sciences of the United States of America, 92*(18), 8135–8139.

Marsh, A. A., Kozak, M. N., Wegner, D. M., Reid, M. E., Yu, H. H., & Blair, R. J. (2010). The neural substrates of action identification. *Social Cognitive and Affective Neuroscience, 5*(4), 392–403.

Martin, A. (2007). The representation of object concepts in the brain. *Annual Review of Psychology, 58*, 25–45.

Maunsell, J. H., & Van Essen, D. C. (1987). Topographic organization of the middle temporal visual area in the macaque monkey: Representational biases and the relationship to callosal connections and myeloarchitectonic boundaries. *Journal of Comparative Neurology, 266*(4), 535–555.

Meltzoff, A. N., & Moore, M. K. (1977). Imitation of facial and manual gestures by human neonates. *Science, 198*, 75–78.

Michels, L., Kleiser, R., de Lussanet, M. H. E., Seitz, R. J., & Lappe, M. (2009). Brain activity for peripheral biological motion in the posterior superior temporal gyrus and the

fusiform gyrus: Dependence on visual hemifield and view orientation. *NeuroImage*, *45*(1), 151–159.
Michels, L., Lappe, M., & Vaina, L. M. (2005). Visual areas involved in the perception of human movement from dynamic form analysis. *Neuroreport*, *16*(10), 1037–1041.
Minnebusch, D. A., & Daum, I. (2009). Neuropsychological mechanisms of visual face and body perception. *Neuroscience and Biobehavioral Reviews*, *33*(7), 1133–1144.
Minnebusch, D. A., Suchan, B., & Daum, I. (2009). Losing your head: Behavioral and electrophysiological effects of body inversion. *Journal of Cognitive Neuroscience*, *21*(5), 865–874.
Mohr, C., Porter, G., & Benton, C. P. (2007). Psychophysics reveals a right hemispheric contribution to body image distortions in women but not men. *Neuropsychologia*, *45*(13), 2942–2950.
Mohr, H. M., Röder, C., Zimmermann, J., Hummel, D., Negele, A., & Grabhorn, R. (2011). Body image distortions in bulimia nervosa: Investigating body size overestimation and body size satisfaction by fMRI. *NeuroImage*, *56*, 1822–1831.
Molenberghs, P., Brander, C., Mattingley, J. B., & Cunnington, R. (2010). The role of the superior temporal sulcus and the mirror neuron system in imitation. *Human Brain Mapping*, *31*(9), 1316–1326.
Mooney, C. M. (1957). Age in the development of closure ability in children. *Canadian Journal of Psychology*, *11*(4), 219–226.
Moro, V., Urgesi, C., Pernigo, S., Lanteri, P., Pazzaglia, M., & Aglioti, S. M. (2008). The neural basis of body form and body action agnosia. *Neuron*, *60*(2), 235–246.
Morrison, I., & Downing, P. E. (2007). Organization of felt and seen pain responses in anterior cingulate cortex. *NeuroImage*, *37*(2), 642–651.
Morrison, I., Lloyd, D., di Pellegrino, G., & Roberts, N. (2004). Vicarious responses to pain in anterior cingulate cortex: Is empathy a multisensory issue? *Cognitive, Affective, & Behavioral Neuroscience*, *4*(2), 270–278.
Morrison, I., Peelen, M. V., & Downing, P. E. (2007). The sight of others'c pain modulates motor processing in human cingulate cortex. *Cerebral Cortex*, *17*(9), 2214–2222.
Morris, J. P., Pelphrey, K. A., & McCarthy, G. (2006). Occipitotemporal activation evoked by the perception of human bodies is modulated by the presence or absence of the face. *Neuropsychologia*, *44*(10), 1919–1927.
Morris, J. S., Friston, K. J., Buchel, C., Frith, C. D., Young, A. W., Calder, A. J., et al. (1998). A neuromodulatory role for the human amygdala in processing emotional facial expressions. *Brain*, *121*(Pt 1), 47–57.
Mouras, H., Stoleru, S., Bittoun, J., Glutron, D., Pelegrini-aIssac, M., Paradis, A. L., et al. (2003). Brain processing of visual sexual stimuli in healthy men: A functional magnetic resonance imaging study. *NeuroImage*, *20*(2), 855–869.
Mumford, D. (1992). On the computational architecture of the neocortex. II. The role of cortico-cortical loops. *Biological Cybernetics*, *66*, 241–251.
Mur, M., Bandettini, P. A., & Kriegeskorte, N. (2009). Revealing representational content with pattern-information fMRI–an introductory guide. *Social Cognitive and Affective Neuroscience*, *4*(1), 101–109.
Myers, A., & Sowden, P. T. (2008). Your hand or mine? The extrastriate body area. *NeuroImage*, *42*(4), 1669–1677.
Newman-Norlund, R. D., Noordzij, M. L., Meulenbroek, R. G. J., & Bekkering, H. (2007). Exploring the brain basis of joint action: Co-ordination of actions, goals and intentions. *Social Neuroscience*, *2*(1), 48–65.
Nummenmaa, L., Hirvonen, J., Parkkola, R., & Hietanen, J. K. (2008). Is emotional contagion special? An fMRI study on neural systems for affective and cognitive empathy. *NeuroImage*, *43*(3), 571–580.
O'Craven, K. M., Downing, P. E., & Kanwisher, N. (1999). fMRI evidence for objects as the units of attentional selection. *Nature*, *401*(6753), 584–587.
O'Craven, K. M., Rosen, B. R., Kwong, K. K., Treisman, A., & Savoy, R. L. (1997). Voluntary attention modulates fMRI activity in human MT-MST. *Neuron*, *18*(4), 591–598.
Oosterhof, N. N., Wiggett, A. J., Diedrichsen, J., Tipper, S. P., & Downing, P. E. (2010). Surface-based information mapping reveals crossmodal vision-action representations in human parietal and occipitotemporal cortex. *Journal of Neurophysiology*, *104*(2), 1077–1089.
Op de Beeck, H. P., Brants, M., Baeck, A., & Wagemans, J. (2010). Distributed subordinate specificity for bodies, faces, and buildings in human ventral visual cortex. *NeuroImage*, *49*(4), 3414–3425.
Oram, M. W., & Perrett, D. I. (1996). Integration of form and motion in the anterior superior temporal polysensory area (STPa) of the macaque monkey. *Journal of Neurophysiology*, *76*(1), 109–129.
Orlov, T., Makin, T. R., & Zohary, E. (2010). Topographic representation of the human body in the occipitotemporal cortex. *Neuron*, *68*(3), 586–600.
Pascual-Leone, A., & Walsh, V. (2001). Fast backprojections from the motion to the primary visual area necessary for visual awareness. *Science*, *2927*, 510–512.
Peelen, M. V., Atkinson, A. P., Andersson, F., & Vuilleumier, P. (2007). Emotional modulation of body-selective visual areas. *Social Cognitive and Affective Neuroscience*, *2*(4), 274–283.
Peelen, M. V., Atkinson, A. P., & Vuilleumier, P. (2010). Supramodal representations of perceived emotions in the human brain. *Journal of Neuroscience*, *30*(30), 10127–10134.
Peelen, M. V., & Downing, P. E. (2005a). Selectivity for the human body in the fusiform gyrus. *Journal of Neurophysiology*, *93*(1), 603–608.
Peelen, M. V., & Downing, P. E. (2005b). Is the extrastriate body area involved in motor actions? *Nature Neuroscience*, *8*(2), 125; author reply 125–125; author reply 126.
Peelen, M. V., & Downing, P. E. (2007a). The neural basis of visual body perception. *Nature Reviews Neuroscience*, *8*(8), 636–648.
Peelen, M. V., & Downing, P. E. (2007b). Using multi-voxel pattern analysis of fMRI data to interpret overlapping functional activations. *Trends in Cognitive Sciences*, *11*(1), 4–5.

Peelen, M. V., Fei-Fei, L., & Kastner, S. (2009). Neural mechanisms of rapid natural scene categorization in human visual cortex. *Nature, 460*(7251), 94–97.

Peelen, M. V., Glaser, B., Vuilleumier, P., & Eliez, S. (2009). Differential development of selectivity for faces and bodies in the fusiform gyrus. *Developmental Science, 12*(6), F16–F25.

Peelen, M. V., Wiggett, A. J., & Downing, P. E. (2006). Patterns of fMRI activity dissociate overlapping functional brain areas that respond to biological motion. *Neuron, 49*(6), 815–822.

Peigneux, P., Van der Linden, M., Garraux, G., Laureys, S., Degueldre, C., Aerts, J., et al. (2004). Imaging a cognitive model of apraxia: The neural substrate of gesture-specific cognitive processes. *Human Brain Mapping, 21*(3), 119–142.

Peuskens, H., Vanrie, J., Verfaillie, K., & Orban, G. A. (2005). Specificity of regions processing biological motion. *European Journal of Neuroscience, 21*(10), 2864–2875.

Piefke, M., Kramer, K., Korte, M., Schulte-Ruther, M., Korte, J. M., Wohischlager, A. M., et al. (2009). Neurofunctional modulation of brain regions by distinct forms of motor cognition and movement features. *Human Brain Mapping, 30*(2), 432–451.

Pierno, A. C., Tubaldi, F., Turella, L., Grossi, P., Barachino, L., Gallo, P., et al. (2009). Neurofunctional modulation of brain regions by the observation of pointing and grasping actions. *Cerebral Cortex, 19*(2), 367–374.

Pietrini, P., Furey, M. L., Ricciardi, E., Gobbini, M. I., Wu, W. H., Cohen, L., et al. (2004). Beyond sensory images: Object-based representation in the human ventral pathway. *Proceedings of the National Academy of Sciences of the United States of America, 101*(15), 5658–5663.

Pinsk, M. A., Arcaro, M., Weiner, K. S., Kalkus, J. F., Inati, S. J., Gross, C. G., & Kastner, S. (2009). Neural representations of faces and body parts in macaque and human cortex: A comparative fMRI study. *Journal of Neurophysiology, 101*, 2581–2600.

Pinsk, M. A., DeSimone, K., Moore, T., Gross, C. G., & Kastner, S. (2005). Representations of faces and body parts in macaque temporal cortex: A functional MRI study. *Proceedings of the National Academy of Sciences of the United States of America, 102*(19), 6996–7001.

Pitcher, D., Charles, L., Devlin, J. T., Walsh, V., & Duchaine, B. (2009). Triple dissociation of faces, bodies, and objects in extrastriate cortex. *Current Biology, 19*, 319–324.

Ponseti, J., Bosinski, H. A., Wolff, S., Peller, M., Jansen, O., Mehdorn, H. M., et al. (2006). A functional endophenotype for sexual orientation in humans. *NeuroImage, 33*(3), 825–833.

Pourtois, G., Peelen, M. V., Spinelli, L., Seeck, M., & Vuilleumier, P. (2007). Direct intracranial recording of body-selective responses in human extrastriate visual cortex. *Neuropsychologia, 45*(11), 2621–2625.

Puce, A., Allison, T., Asgari, M., Gore, J. C., & McCarthy, G. (1996). Differential sensitivity of human visual cortex to faces, letter strings, and textures: A functional magnetic resonance imaging study. *Journal of Neuroscience, 16*(16), 5205–5215.

Puce, A., & Perrett, D. (2003). Electrophysiology and brain imaging of biological motion. *Philosophical Transactions of the Royal Society of London. Series B, Biological Sciences, 358*(1431), 435–445.

Quadflieg, S., Flannigan, N., Waiter, G. D., Rossion, B., Wig, S. G., Turk, D. J., & Macrae, C. N. (2011). Stereotype-based modulation of person perception. *NeuroImage, 57*, 549–557.

Ramsey, R., & Hamilton, A. F. (2010). Understanding actors and object-goals in the human brain. *NeuroImage, 50*(3), 142–1147.

Reid, V. M., Hoehl, S., & Striano, T. (2006). The perception of biological motion by infants: An event-related potential study. *Neuroscience Letters, 395*, 211–214.

Renier, L. A., Anurova, I., De Volder, A. G., Carlson, S., Van Meter, J., & Rauschecker, J. P. (2010). Preserved functional specialization for spatial processing in the middle occipital gyrus of the early blind. *Neuron, 68*, 138–148.

Ricciardi, E., Bonino, D., Sani, L., Vecchi, T., Guazzelli, M., Haxby, J. V., et al. (2009). Do we really need vision? How blind people "see" the actions of others. *Journal of Neuroscience, 29*(31), 9719–9724.

Rossion, B., Caldara, R., Seghier, M., Schuller, A. M., Lazeyras, F., & Mayer, E. (2003). A network of occipito-temporal face-sensitive areas besides the right middle fusiform gyrus is necessary for normal face processing. *Brain, 126*(Pt 11), 2381–2395.

Rueschemeyer, S. A., Pfeiffer, C., & Bekkering, H. (2010). Body schematics: On the role of the body schema in embodied lexical-semantic representations. *Neuropsychologia, 48*(3), 774–781.

Sachdev, P., Mondraty, N., Wen, W., & Gulliford, K. (2008). Brains of anorexia nervosa patients process self-images differently from non-self-images: An fMRI study. *Neuropsychologia, 46*(8), 2161–2168.

Saxe, R., Jamal, N., & Powell, L. (2006). My body or yours? The effect of visual perspective on cortical body representations. *Cerebral Cortex, 16*(2), 178–182.

Saxe, R., Xiao, D. K., Kovacs, G., Perrett, D. I., & Kanwisher, N. (2004). A region of right posterior superior temporal sulcus responds to observed intentional actions. *Neuropsychologia, 42*(11), 1435–1446.

Schendan, H. E., & Kutas, M. (2003). Time course of processes and representations supporting visual object identification and memory. *Journal of Cognitive Neuroscience, 15*(1), 111–135.

Schiller, P. H., Finlay, B. L., & Volman, S. F. (1976). Quantitative studies of single-cell properties in monkey striate cortex. I. Spatiotemporal organization of receptive fields. *Journal of Neurophysiology, 397*, 1288–1319.

Schwarzlose, R. F., Baker, C. I., & Kanwisher, N. (2005). Separate face and body selectivity on the fusiform gyrus. *Journal of Neuroscience, 25*(47), 11055–11059.

Schwarzlose, R. F., Swisher, J. D., Dang, S. B., & Kanwisher, N. (2008). The distribution of category and location information across object-selective regions in human visual cortex. *Proceedings of the National Academy of Sciences of the United States of America, 105*(11), 4447–4452.

Schwoebel, J., & Coslett, H. B. (2005). Evidence for multiple, distinct representations of the human body. *Journal of Cognitive Neuroscience*, *17*(4), 543–553.

Singer, J. M., & Sheinberg, D. L. (2010). Temporal cortex neurons encode articulated actions as slow sequences of integrated poses. *Journal of Neuroscience*, *30*(8), 3133–3145.

Singer, T., Seymour, B., O'Doherty, J., Kaube, H., Dolan, R. J., & Frith, C. D. (2004). Empathy for pain involves the affective but not sensory components of pain. *Science*, *303*(5661), 1157–1162.

Slaughter, V., & Heron, M. (2004). Origins and early development of human body knowledge. *Monographs of the Society for Research in Child Development*, *69*(2), vii, 1–102.

Southgate, V., Johnson, M. H., & Csibra, G. (2008). Infants attribute goals even to biomechanically impossible actions. *Cognition*, *107*, 1059–1069.

Spiridon, M., Fischl, B., & Kanwisher, N. (2006). Location and spatial profile of category-specific regions in human extrastriate cortex. *Human Brain Mapping*, *27*(1), 77–89.

Spiridon, M., & Kanwisher, N. (2002). How distributed is visual category information in human occipito-temporal cortex? An fMRI study. *Neuron*, *35*(6), 1157–1165.

Spunt, R. P., Falk, E. B., & Lieberman, M. D. (2010). Dissociable neural systems support retrieval of how and why action knowledge. *Psychological Science*, *21*(11), 1593–1598.

Spunt, R. P., Satpute, A. B., & Lieberman, M. D. (2011). Identifying the what, why, and how of an observed action: An fMRI study of mentalizing and mechanizing during action observation. *Journal of Cognitive Neuroscience*, *23*(1), 63–74.

Suchan, B., Busch, M., Schulte, D., Gronermeyer, D., Herpertz, S., & Vocks, S. (2010). Reduction of gray matter density in the extrastriate body area in women with anorexia nervosa. *Behavioural Brain Research*, *206*(1), 63–67.

Sugita, Y. (2009). Innate face processing. *Current Opinion in Neurobiology*, *19*, 39–44.

Sugiura, M., Sassa, Y., Jeong, H., Miura, N., Akitsuki, Y., Horie, K., et al. (2006). Multiple brain networks for visual self-recognition with different sensitivity for motion and body part. *NeuroImage*, *32*(4), 1905–1917.

Summerfield, C., Trittschuh, E. H., Monti, J. M., Mesulam, M. M., & Egner, T. (2008). Neural repetition suppression reflects fulfilled perceptual expectations. *Nature Neuroscience*, *11*(9), 1004–1006.

Takahashi, H., Shibuya, T., Kato, M., Sassa, T., Koeda, M., Yahata, N. et al. (2008). Enhanced activation in the extrastriate body area by goal-directed actions. *Psychiatry and Clinical Neurosciences*, *62*(2), 214–219.

Taylor, J. C., & Downing, P. E. (2011). Division of labor between lateral and ventral extrastriate representations of faces, bodies, and objects. *Journal of Cognitive Neuroscience*. Advance online publication. doi: 10.1162/jocn_a_00091.

Taylor, J. C., Roberts, M. V., Downing, P. E., & Thierry, G. (2010). Functional characterisation of the extrastriate body area based on the N1 ERP component. *Brain and Cognition*, *73*(3), 153–159.

Taylor, J. C., Wiggett, A. J., & Downing, P. E. (2007). Functional MRI analysis of body and body part representations in the extrastriate and fusiform body areas. *Journal of Neurophysiology*, *98*(3), 1626–1633.

Taylor, J. C., Wiggett, A. J., & Downing, P. E. (2010). fMRI-adaptation studies of viewpoint tuning in the extrastriate and fusiform body areas. *Journal of Neurophysiology*, *103*(3), 1467–1477.

Thierry, G., Pegna, A. J., Dodds, C., Roberts, M., Basan, S., & Downing, P. (2006). An event-related potential component sensitive to images of the human body. *NeuroImage*, *32*(2), 871–879.

Tsao, D. Y., Freiwald, W. A., Knutsen, T. A., Mandeville, J. B., & Tootell, R. B. (2003). Faces and objects in macaque cerebral cortex. *Nature Neuroscience*, *6*(9), 989–995.

Uher, R., Murphy, T., Friederich, H. C., Dalgleish, T., Brammer, M. J., Giampietro, V., et al. (2005). Functional neuroanatomy of body shape perception in healthy and eating-disordered women. *Biological Psychiatry*, *58*(12), 990–997.

Umilta, M. A., Escola, L., Intskirveli, I., Grammont, F., Rochat, M., Caruana, F., et al. (2008). When pliers become fingers in the monkey motor system. *Proceedings of the National Academy of Sciences of the United States of America*, *105*(6), 2209–2213.

Urgesi, C., Berlucchi, G., & Aglioti, S. M. (2004). Magnetic stimulation of extrastriate body area impairs visual processing of nonfacial body parts. *Current Biology*, *14*(23), 2130–2134.

Urgesi, C., Calvo-Merino, B., Haggard, P., & Aglioti, S. M. (2007). Transcranial magnetic stimulation reveals two cortical pathways for visual body processing. *Journal of Neuroscience*, *27*, 8023–8030.

Urgesi, C., Candidi, M., Ionta, S., & Aglioti, S. M. (2007). Representation of body identity and body actions in extrastriate body area and ventral premotor cortex. *Nature Neuroscience*, *10*, 30–31.

Valyear, K. F., & Culham, J. C. (2010). Observing learned object-specific functional grasps preferentially activates the ventral stream. *Journal of Cognitive Neuroscience*, *22*(5), 970–984.

Van Dam, W. O., Rueschemeyer, S. A., & Bekkering, H. (2010). How specifically are action verbs represented in the neural motor system: An fMRI study. *NeuroImage*, *53*(4), 1318–1325.

Van de Riet, W. A. C., Grezes, J., & de Gelder, B. (2009). Specific and common brain regions involved in the perception of faces and bodies and the representation of their emotional expressions. *Social Neuroscience*, *4*(2), 101–120.

Van Overwalle, F., & Baetens, K. (2009). Understanding others' actions and goals by mirror and mentalizing systems: A meta-analysis. *Neuroimage*, *48*(3), 564–584.

Vocks, S., Busch, M., Gronemeyer, D., Schulte, D., Herpertz, S., & Suchan, B. (2010). Differential neuronal responses to the self and others in the extrastriate body area and the fusiform body area. *Cognitive, Affective, & Behavioral Neuroscience*, *10*(3), 422–429.

Vuilleumier, P. (2005). How brains beware: Neural mechanisms of emotional attention. *Trends in Cognitive Sciences*, *9*(12), 585–594.

Vuilleumier, P., Richardson, M. P., Armony, J. L., Driver, J., & Dolan, R. J. (2004). Distant influences of amygdala lesion on visual cortical activation during emotional face processing. *Nature Neuroscience*, *7*(11), 1271–1278.

Wachsmuth, E., Oram, M. W., & Perrett, D. I. (1994). Recognition of objects and their component parts: Responses of single units in the temporal cortex of the macaque. *Cerebral Cortex*, *4*(5), 509–522.

Weiner, K. S., & Grill-Spector, K. (2010). Sparsely-distributed organization of face and limb activations in human ventral temporal cortex. *NeuroImage*, *52*(4), 1559–1573.

Weiner, K. S., & Grill-Spector, K. (2011). Not one extrastriate body area: Using anatomical landmarks, hMT+, and visual field maps to parcellate limb-selective activations in human lateral occipitotemporal cortex. *NeuroImage*, *56*(4), 2183–2199.

Whitney, D., Ellison, A., Rice, N. J., Arnold, D., Goodale, M., Walsh, V., et al. (2007). Visually guided reaching depends on motion area MT+. *Cerebral Cortex*, *17*(11), 2644–2649.

Wiggett, A. J., & Downing, P. E. (2008). The face network: Overextended? (Comment on: "Let's face it: It's a cortical network" by Alumit Ishai). *NeuroImage*, *40*(2), 420–422.

Wiggett, A. J., & Downing, P. E. (2011). Representation of action in occipito-temporal cortex. *Journal of Cognitive Neuroscience*, *23*(7), 1765–1780.

Willems, R. M., Peelen, M. V., & Hagoort, P. (2010). Cerebral lateralization of face-selective and body-selective visual areas depends on handedness. *Cerebral Cortex*, *20*(7), 1719–1725.

Wojciulik, E., Kanwisher, N., & Driver, J. (1998). Covert visual attention modulates face-specific activity in the human fusiform gyrus: fMRI study. *Journal of Neurophysiology*, *79*(3), 1574–1578.

Yomogida, Y., Sugiura, M., Sassa, Y., Wakusawa, K., Sekiguchi, A., Fukushima, A., et al. (2010). The neural basis of agency: An fMRI study. *NeuroImage*, *50*(1), 198–207.

Yovel, G., Pelc, T., & Lubetzky, I. (2010). It's all in your head: Why is the body inversion effect abolished for headless bodies? *Journal of Experimental Psychology: Human Perception and Performance*, *36*(3), 759–767.

Zimmermann, M., Meulenbroek, R. G., & de Lange, F. P. (2011). Motor planning is facilitated by adopting an action's goal posture: An fMRI study. *Cerebral Cortex*. Advance online publication. doi: 10.1093/cercor/bhr098.

COGNITIVE NEUROSCIENCE, 2011, 2 (3–4), 227–257 (includes Discussion Paper, Commentaries, and Reply)

Discussion Paper

On the functional anatomy of the urge-for-action

Stephen R. Jackson[1,2], Amy Parkinson[2], So Young Kim[1], Martin Schüermann[2], and Simon B. Eickhoff[3,4]

[1]Department of Brain and Cognitive Engineering, Korea University, Seoul, South Korea
[2]School of Psychology, University of Nottingham, Nottingham, UK
[3]Department of Psychiatry and Psychotherapy, RWTH Aachen University, Aachen, Germany
[4]Institute of Neuroscience and Medicine (INM – 2), Research Center Jülich, Jülich, Germany

Several common neuropsychiatric disorders (e.g., obsessive-compulsive disorder, Tourette syndrome (TS), autistic spectrum disorder) are associated with unpleasant bodily sensations that are perceived as an urge for action. Similarly, many of our everyday behaviors are also characterized by bodily sensations that we experience as urges for action. Where do these urges originate? In this paper, we consider the nature and the functional anatomy of "urges-for-action," both in the context of everyday behaviors such as yawning, swallowing, and micturition, and in relation to clinical disorders in which the urge-for-action is considered pathological and substantially interferes with activities of daily living (e.g., TS). We review previous frameworks for thinking about behavioral urges and demonstrate that there is considerable overlap between the functional anatomy of urges associated with everyday behaviors such as swallowing, yawning, and micturition, and those urges associated with the generation of tics in TS. Specifically, we show that the limbic sensory and motor regions—insula and mid-cingulate cortex—are common to all of these behaviors, and we argue that this "motivation-for-action" network should be considered distinct from an "intentional action" network, associated with regions of premotor and parietal cortex, which may be responsible for the perception of "willed intention" during the execution of goal-directed actions.

Keywords: Urge; Urge-for-action; Habit; Insula; Tourette syndrome; Action.

Many of our everyday behaviors are characterized by bodily sensations that we experience either as an *urge* or a *desire* for action. For instance, we may experience a sensation that our bladder is full that is accompanied, to a greater or lesser extent, by an urge or desire to urinate (micturate). In extreme cases, this sense of fullness can be quite uncomfortable and the urge to urinate can be hard to suppress. Similarly, we may experience a tickle in our throat that is associated with an urge to cough or to swallow that can also be difficult to suppress voluntarily. However, not all urges for action are necessarily preceded by bodily sensations of which we are aware. For example, we may suddenly experience a strong urge to yawn, or even find ourselves yawning, without being aware of a sensory "trigger" for the action. In this paper, we consider the nature and the functional anatomy of these "urges-for-action," both in the context of everyday behaviors such as yawning, swallowing, and urinating, and in relation to clinical disorders in which the urge-for-action is considered pathological and substantially

Correspondence should be addressed to: Stephen R. Jackson, School of Psychology, University of Nottingham, Nottingham NG7 2RD, UK. E-mail: stephen.jackson@nottingham.ac.uk

This research was supported in part by grants from the Medical Research Council (G0901321), and by the WCU (World Class University) program through the National Research Foundation of South Korea, and funded by the Ministry of Education, Science, and Technology (R31-10008). We are particularly grateful to two anonymous referees for their helpful comments. SBE acknowledges funding by the Human Brain Project (R01-MH074457-01A1), the DFG (IRTG 1328) and the Helmholtz Initiative on Systems-Biology.

http://dx.doi.org/10.1080/17588928.2011.604717

interferes with activities of daily living (e.g., Tourette syndrome, obsessive-compulsive disorder, addiction).

A CONCEPTUAL MODEL OF THE URGE-FOR-ACTION

In common usage, the terms "urge" and "desire" are frequently encountered as both a verb (as in to "*urge* someone on"—i.e., to motivate, impel, or stimulate a person toward an action) and as a noun (as in "he felt an *urge* to shout"—i.e., a force, drive, or impulse that impels toward a goal). Furthermore, these terms are often used interchangeably with one another, and are frequently listed as synonyms.

Nevertheless, it has been suggested that there is an important distinction to be made between urges and desires (e.g., Cameron, 2002; Davenport, 2008). Thus, Cameron (2002), when discussing interoception, and the relationship between conscious awareness and visceral events, makes a distinction between "detection," which is an organism's reflexive response based solely upon afferent physiological information, and "perception," which refers to an organism's response based upon all information available to the organism (which might include learned information and expectations that might be generated as a result of learning). Similarly, Davenport, when discussing mechanisms associated with the urge-to-cough, defines an *urge* as a physical need to respond to a sensory stimulus, and a *desire* as the translation of an urge into what he refers to as "a central neural targeted goal" (Davenport, Sapienza, & Bolser, 2002). Davenport's motivation-to-action model is presented in Figure 1.

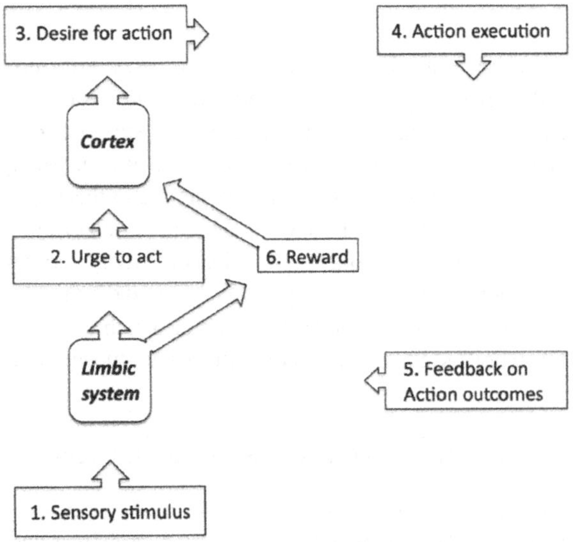

Figure 1. The motivation-for-action model proposed by Davenport to account for the urge to cough (adapted from Davenport, Sapienza, & Bolser, 2002).

Combining these two ideas we might conclude that an urge—as in a drive for action—need not enter conscious awareness, but that we are always aware of our desires. An example might help make this distinction clearer. Imagine that you are driving your car and you suddenly become aware of an uncomfortable sensation that your bladder is full, which you experience as the *urge to urinate*. Given that you can in all likelihood control this urge, you may then form a *desire to urinate* in which you construct a goal or plan which includes a representation of the behaviors required to complete the action (e.g., remembering that you passed a filling station a mile back and knowing that filling stations usually have public toilets) and a representation of the likely outcomes of the action. Can an urge exist if we are not aware of it? What factors determine whether an urge enters awareness? How is an urge to act different from an intention to act? These issues are discussed below.

ARE URGES SIMPLY REFLEXES?

As outlined above, urges are often defined as the drives or impulses that impel us to act. Nevertheless, it is argued that actions can, and frequently do, occur in the absence of any awareness of such drivers, as in the case where one finds oneself yawning without previously being aware of either the desire to yawn or of any bodily sensation that might reasonably be identified as giving rise to the yawn. Instead, one simply finds oneself yawning. In this case, it might be argued that this is a *reflexive* behavior and not an urge-for-action.

One possible distinguishing feature of urges, as distinct from reflexes, may be that urges are chiefly associated with actions that cannot be realized immediately and must be held in check until an appropriate time when they might be released. For instance, when we become aware of having a full bladder, we experience an "urge-to-void" because we do not simply void our bladder, but instead employ a coordinated set of central, autonomic, and peripheral neural mechanisms to withhold micturition until we are in an appropriate behavioral context. Similarly, in the case of yawning, we might define the urge-to-yawn as arising in circumstances where we are forced to try to stifle the yawn rather than in the situation where we find that we are yawning.

HOW ARE URGES RELATED TO AWARENESS?

One factor that may determine whether an urge enters awareness is the intensity of the physiological afferent.

It has been demonstrated in the context of the urge-to-cough that the perceived strength of the urge that is experienced is related to the intensity of stimulation. Specifically, when capsaicin is added to the breathing circuit, it results in a sensation that is perceived as an urge-to-cough. Furthermore, increasing capsaicin levels leads to a reliable increase in estimates of this urge (Davenport, Sapienza, & Bolser, 2002). Similarly, in our own unpublished studies of the effects of oropharyngeal stimulation on the urge-to-swallow, we have found that oropharyngeal stimulation using pulses of air produces both an urge-to-swallow and overt swallowing (for similar findings, see also Lowell et al., 2008). More importantly, we have found that increasing the intensity of oropharyngeal stimulation leads to an increase in the strength of the perceived urge-to-swallow. An important question therefore concerns the role of awareness, and by implication "desire," in the initiation of urge-related actions.

According to the motivation-to-action model represented in Figure 1, actions are initiated only after a desire to perform an action has been formed. However, we do not see this step as necessary and would amend this model, so as to allow for actions to be initiated directly, without the need to first construct an explicit "desire for action." Two factors motivate this amendment. First, it strikes us that in many instances actions (e.g., yawning) can be triggered without our necessarily being aware of any explicit desire. Second, in many clinical conditions (e.g., Tourette syndrome), as we shall see below, *unwanted* actions or behaviors, which individuals struggle actively to suppress, are nevertheless preceded by strong premonitionary urges. In our view, it is difficult to reconcile the concept of a "desire," which is often defined as "a longing or craving for something that brings satisfaction or enjoyment," with unwanted actions, the execution of which is experienced as unpleasant and distressing.

If we accept the premise outlined above that urges-for-action are often accompanied by bodily sensations, then it strikes us that an important distinction can be drawn between being aware of a bodily sensation and being aware of an urge-for-action. This distinction can be best illustrated by considering the following examples. In the case of an itch, we may be aware of experiencing both an itch (bodily sensation) and an urge to scratch the itch (urge-for-action). By contrast, while we may become aware of an urge-to-yawn, it is not entirely clear that we are ever aware of the bodily sensation that gives rise to the urge to yawn. Furthermore, this becomes more important if we accept the argument outlined above that urges occur primarily in circumstances in which actions may need to be suppressed or their execution deferred. In such circumstances, we might distinguish between suppression of the *action* associated with the urge-for-action, or suppression of the *bodily sensation* giving rise to the urge. This distinction has important clinical implications (see below), and an interesting issue for future research will be to determine whether the suppression of bodily sensations and the suppression of urges-for-action differ in terms of their functional anatomy.

HOW IS AN URGE TO ACT DIFFERENT FROM AN INTENTION TO ACT?

The kinds of actions that we have considered as representative of urges-for-action are highly automatic, habitual responses that occur primarily in response to sensory stimulation. These might include brushing an insect off your arm, scratching an itch, yawning when tired, coughing in response to a tickle in your throat, etc. While such actions can, in some circumstances, be executed with little or no awareness of the sensory stimulation that triggered the action, as when one finds oneself yawning or coughing, we have argued that a key characteristic of urges-for-action is that they involve the suppression or deferment of an action. Such actions might therefore be contrasted to intentional, goal-directed, forms of action.

The circumstances in which the "willed intention" to execute an action can be shown to follow the brain processes involved in the preparation for action were famously studied by Benjamin Libet (Libet, Gleason, Wright, & Pearl, 1983), and more recently by Patrick Haggard and colleagues (e.g., Haggard, 2005; Haggard & Eimer, 1999; Sirigu et al., 2004). In Libet's task, participants fixated on a time-varying, rotating visual spot and were instructed to make a voluntary hand movement whenever they felt the "urge" to do so. Participants were asked to indicate the location occupied by the moving spot when they had first felt the urge to move their hand. Libet showed that this "W judgment" occurred some 200 ms prior to movement onset, but, more importantly, he also showed that the preparatory brain activity that precedes voluntary action, the so-called "readiness potential," itself preceded the "W judgment" by several hundred milliseconds. Such readiness potentials arise in the premotor regions of cortex, including both the supplementary motor area (SMA) and the presupplementary motor area (pre-SMA), regions that have been linked to the planning and preparation of intentional, goal-directed, actions and sequences of actions (Deecke & Kornhuber, 1978).

Haggard has argued that conscious awareness of our intention to act arises during the preparatory processes that precede an action, and is linked to the joint activity of premotor and parietal brain areas (Haggard, 2005).

In support of this view, he has shown that patients with damage to the parietal cortex show a specific impairment in reporting when they became aware of their intention to move (i.e., Libet's "W judgment"). The proposal that the parietal cortex may maintain a dynamically updated state estimate of the current postural configuration of the body (the body schema) is well supported by neuropsychological (e.g., Wolpert, Goodbody, & Husain, 1998) and recent fMRI studies (e.g., Parkinson, Condon, & Jackson, 2010; Pellijeff, Bonilha, Morgan, McKenzie, & Jackson, 2006).

While it is clearly the case that in such experiments individuals can report when they first perceived themselves to have formed an "intention" to move, it is another thing entirely to argue that such conscious "intentions" typically precede everyday actions. Thus, when I am sitting at my desk typing and I break off to reach for my coffee cup I am not aware of forming an intention prior to each keystroke that I make or of forming an intention to reach for my coffee. Instead, I am aware of the actions I am making. Similarly, while it is clearly demonstrated that our actions are preceded by neural activity, as discussed by Libet and by Haggard and colleagues, it is currently unclear how these activations relate to the phenomenology of intention.

Haggard and others (e.g., Blakemore, Wolpert, & Frith, 2002; Haggard, 2005) have argued that the sense-of-agency that typically accompanies the execution of voluntary movements arises as a result of internal forward sensory models that generate a prediction of the sensory consequences of an action that is then matched against afferent sensory signals. It has been proposed that in cases where the link between these sensory predictions and confirmatory sensory input is broken, neurological syndromes such as anosagnosia (lack of awareness of injury) or somatoparaphrenia (denial of limb ownership) may occur (Tsakiris, 2010). Interestingly, both of these disorders have been linked to damage of the anterior insular cortex of the right hemisphere (Baier & Karnath, 2008; Karnath, Baier, & Nagele, 2005). However, it is important to note that this sense of agency may in fact have a significant postdictive or reconstructive component (Moore, Lagnado, Deal, & Haggard, 2009; Wegner, 2002) and thus is not necessarily an unambiguous index of intentionality.

URGES-FOR-ACTION, TICS, AND TOURETTE SYNDROME

Tics are involuntary, repetitive, stereotyped behaviors that occur with a limited duration. Motor tics can be simple or complex in appearance, ranging from repetitive movements to coordinated action sequences (Leckman, 2002). Verbal tics can consist of repeating words or utterances (palilalia), producing inappropriate or obscene utterances (coprolalia), or the repetition of another's words (echolalia).

Tics occur in bouts, typically many times in a single day, and are the most common form of movement disorder in children, with a prevalence of 1–29% depending upon the precise characteristics of the study population, the diagnostic criteria used, and the study design and methods employed (Leckman, 2002). Tics include a continuum of disorders: transient tic disorder (TTD), chronic tic disorder (CTD), nonspecific tic disorder (NSTD), and Tourette syndrome (TS). The etiology of tics is poorly understood and probably involves a complex interaction between genetic and environmental factors that exert an influence over brain development.

TS is a developmental neuropsychiatric disorder that lies at the extreme of the tic disorder spectrum and is characterized by the presence of chronic vocal and motor tics (Leckman, 2002). The neurological basis of TS is unclear; however, it is agreed that the basal ganglia, including circuits that link the striatum to the frontal lobes, are dysfunctional (Albin & Mink, 2006). A specific model of basal ganglia dysregulation in TS has been proposed as follows. Subsets of striatal neurons (matrisomes) are thought to become abnormally active in inappropriate contexts, leading to the disinhibition of thalamocortical circuits that in turn lead to tics. Activity-dependent dopamine inappropriately reinforces such activity, leading to stereotyped repetition of behavior (Albin & Mink, 2006). Brain-imaging and postmortem studies provide general support for the view that cortical–striatal–thalamocortical pathways are dysfunctional in TS (Gerard & Peterson, 2003). Furthermore, deep-brain stimulation of the globus pallidus or the thalamus has been shown to be effective in suppressing tics in individuals with TS (e.g., Ackermans et al., 2011).

The occurrence of repetitive, stereotypical behaviors in TS has been linked to operation of the brain "reward" and "habit" systems, and tics have been likened to an inappropriate overextension of habit learning (Graybiel, 2008). Key characteristics of habitual behaviors are that they are largely learnt, occur repeatedly, are performed almost automatically, and often involve stereotypical, ordered, action sequences (Graybiel, 2008). In this context, it is important to note that many individuals with TS report that their tics are often preceded by "premonitory sensory phenomena" (PSPs), which are described as the presence of uncomfortable cognitive or bodily sensations (e.g., tension, pressure, tickle), that precede the execution of a tic, and are experienced as a strong *urge for motor discharge* (Banaschewki, Woerner, & Rothenberger,

2003). Furthermore, whereas individuals with TS perceive a relatively constant demand to suppress their tics in social situations, and while the voluntary suppression of tics is possible in many cases, individuals with TS nevertheless report that it can be uncomfortable and stressful to suppress tics, and that the urge to tic becomes uncontrollable after a period of suppression. It is therefore likely that some tics at least are learnt motor or vocal behaviors that function to alleviate or reduce uncomfortable bodily sensations.

Functional anatomy of tics in TS

Several attempts have been made to investigate the brain regions associated with the occurrence of tics in TS using human neuroimaging techniques. These studies have indicated that the neural mechanisms responsible for triggering of tics may in fact differ from those involved in voluntary movements (Bohlhalter et al. 2006). One notable study was that reported by Hallett and colleagues, which used functional magnetic resonance imaging (fMRI) to examine brain areas activated immediately preceding the spontaneous occurrence of motor and/or vocal tics, and thus likely to be associated with the urge to tic (Bohlhalter et al., 2006). This study identified a network of brain areas that were activated immediately prior to tic onset, and, most importantly, identified the insular cortex, the anterior cingulate cortex, and the parietal operculum, which includes the secondary somatosensory cortex (SII) (Eickhoff, Amunts, Mohlberg, & Zilles, 2006), as the most likely anatomic regions responsible for the uncomfortable feelings associated with premonitory urges to tic (Bohlhalter et al., 2006). Consistent with this proposal, electrical stimulation of the insular cortex or the parietal operculum can elicit unpleasant somatosensory or visceral sensations (Augustine, 1996; Ostrowsky et al., 2002; Penfield & Faulk, 1955). By contrast, electrical stimulation of the medial frontal lobes produces motor outputs in the face and upper limbs comparable to tics (Bancaud et al., 1976; Lim et al., 1994; Talairach et al., 1973).

The proposal that the insular and cingulate cortices are associated with the uncomfortable feelings associated with the urge to tic in TS is consistent with the putative role of these areas in the neural representation of bodily states more generally (interoception), and the initiation of behaviors associated with these bodily representations (for reviews, see Craig, 2002, 2009; Naqvi & Bechara, 2008). Thus, Craig suggests that these two regions are linked functionally and can be thought of as the limbic sensory and motor areas. He has proposed that these two areas form part of a functional brain system that is associated with the *awareness* of bodily states (particularly the right insular cortex) and the maintenance of homeostasis (Craig, 2009; cf. Damasio, 1999).

It is important to note that Craig's particular view of interoception includes a representation of all body states relevant to homeostasis, including pain, temperature, taste, visceral sensation, inflammation, itch, and many aspects of touch and proprioception that are often viewed as part of an "exteroceptive" somatosensory system (Craig, 2003). Consistent with this view, recent functional brain-imaging studies have demonstrated that punctate somatosensory stimulation of the upper limbs produces significant increases in brain activity—blood oxygen level-dependent (BOLD) response)—bilaterally within the insular cortex (e.g., Jackson, Parkinson, Pears, & Nam, 2011; Parkinson et al., 2011). However, neurophysiological studies suggest that the insular cortex may play a particularly important role in representing the emotional significance of somatosensory signals. Thus, it has been shown that a neural pathway, consisting of unmyelinated fibers, projects to the insula (Olausson et al., 2002; Vallbo, Olausson, & Wessberg, 1999), and that these fibers are associated with affective or sensual touch (e.g., pleasant touch sensation).

Brain-imaging and neurological studies also indicate that the posterior and mid-insular cortex may play an important role in body ownership and our sense of agency (control) over our body. Thus, a positron emission tomography (PET) imaging study reported by Tsakiris, Hesse, Boy, Haggard, and Fink (2007) reported that body ownership, as indexed by the strength of the rubber hand illusion, was associated with activation increases within the right posterior insula and right frontal operculum. Similarly, Karnath and colleagues have shown that lesions involving the right insula impair body awareness, and the sense of limb ownership (e.g., Baier & Karnath, 2008; Karnath, Baier, & Nagele, 2005).

Arguably, the most direct neuropsychological evidence that the insular cortex is key to the experience of urges for action comes from a study that investigated the effect of insula lesions on the urge to smoke in those addicted to smoking (Naqvi, Rudrauf, Damasio, & Bechara, 2007). This study compared smokers who had sustained damage involving the insula with a group of smokers whose damage involved other brain areas, but spared the insula. The study investigated changes in smoking behavior post-stroke and demonstrated that smokers whose brain damage involved the insula were significantly more likely than smokers with lesions sparing the insula to exhibit a "disruption of smoking addiction." Importantly, individuals described this disruption of addiction as like their body "forgetting the urge to smoke" (Naqvi et al., 2007).

LIMITATIONS OF INDIVIDUAL BRAIN-IMAGING STUDIES

Functional brain-imaging studies using fMRI have become central to cognitive neuroscience; however, it should be recognized that such studies have limitations. Many of these are well known and relate to the constraints imposed by hemodynamic signals (Logothetis, 2008). A discussion of these limitations is generally beyond the scope of this paper; however, below, we briefly outline some issues that specifically relate to difficulties in interpreting fMRI activations associated with the urge to tic in individuals with TS.

First, one obvious difficulty associated with interpreting the meaning of the patterns of BOLD activity reported in fMRI studies that have sought to identify brain regions associated with the occurrence of tics in TS, is that individuals with TS are instructed to remain still in the MRI scanner and to suppress their tics. Thus, the regions activated can reflect brain areas associated with the generation of tics or brain areas linked to their active suppression. Second, recent evidence suggests that individuals with neurodevelopmental disorders may follow unique developmental trajectories whereby they undergo compensatory, neuroplastic changes in brain structure and function that help them gain control over their symptoms (Jackson, Parkinson, Jung, et al., 2011). As a consequence, individuals with TS may exhibit differences in functional anatomy, compared to typically developing individuals, even when they are performing an identical behavioral task with comparable levels of task performance. Third, the functional anatomy of the urge to tic in TS may differ from the functional anatomy of other forms of the urge for action. In fact, a related limitation of fMRI studies is that it is very often difficult to carry out the range of control studies that might be conducted when using non-imaging experimental techniques. Finally, an important limitation of individual fMRI studies is that it is very often difficult to compare across studies. Thus, a comparison of individual fMRI studies of, for example, the urge to urinate and the urge to swallow, might reveal differences in functional anatomy because these behaviors have different underlying neural circuitry, or because of differences in the following factors: the behavioral paradigms used, the scanner hardware and imaging protocols adopted, analysis protocols and statistical thresholds, etc.

This last limitation can be overcome, however, through the use of quantitative meta-analytic studies. Such studies permit an estimation of the fMRI BOLD response associated with different behaviors by drawing upon the entire body of published studies within a particular behavioral domain. One method that has proven popular recently has been the activation likelihood estimation (ALE) method developed by Turkeltaub, Eden, Jones, and Zeffiro (2002) and modified by Eickhoff and colleagues (Eickhoff et al., 2009).

A QUANTITATIVE META-ANALYTIC COMPARISON OF THE URGE TO MICTURATE AND THE URGE TO SWALLOW

Here we report the use of the ALE method to directly compare brain activity associated with the urge to urinate (micturition) and the urge to swallow.

Study selection criteria

Functional neuroimaging studies were retrieved via searches in the PubMed, ISI Web of Knowledge, and Scopus databases, as well as identified by reference tracing and through reviews. Experiments reported in these papers that corresponded to the core behavioral contrasts under consideration, such as contrast of urine withholding against a resting baseline, urine voiding against baseline, and volitional swallowing against baseline, were included in the meta-analyses if they fulfilled the following criteria: analyses must be computed across the whole brain and not restricted by partial coverage or regions of interest analyses; coordinates must be reported in an XYZ format, either in MNI or Talairach space; only experiments that investigated differences between stimulation conditions in healthy control population were included; and experiments focusing on between-group differences were excluded. No selection was made on the basis of the applied statistical threshold, as all studies were obtained from peer-reviewed journals. Additional methodological information is provided in detail in the online supplementary material, available via the 'Supplementary' tab on the article's online page (http://dx.doi.org/10.1080/17588928.2011.604717).

Experiments investigating the functional anatomy of *swallowing* were selected if they contrasted saliva swallowing, water swallowing, overt swallowing, or covert swallowing with a rest condition or other suitable control condition. Experiments investigating the functional anatomy of *micturition* were selected if they contrasted micturition, urine withholding, bladder filling, or the urge to void with a rest condition or other suitable control condition. See Supplementary Material for further details.

The quantitative comparison of ALEs across studies of swallowing revealed widespread clusters of activation that exceeded the conservative statistical threshold (Figure 2A). The areas of cortex that exceeded threshold included primary motor cortex (BA 4), premotor cortex

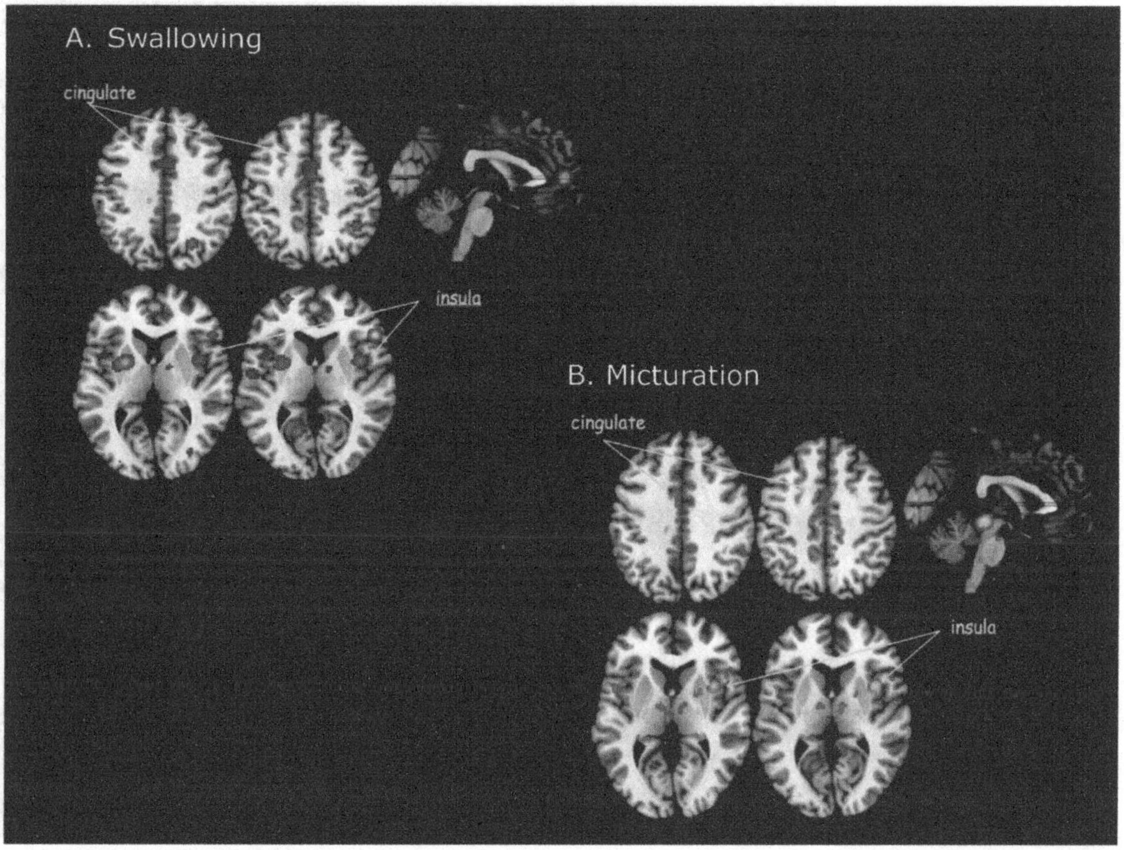

Figure 2. (A) Main results of an ALE meta-analysis of neuroimaging studies of swallowing. This analysis revealed a number of activation foci that survived conservative statistical correction ($p < .05$ corrected for false discovery rate). Among these were activations within the insular cortex and the dorsal mid-cingulate cortex. (B) Main results of an ALE meta-analysis of neuroimaging studies of micturition. Again this analysis revealed statistically significant foci of activation within the insular cortex and the dorsal mid-cingulate cortex.

(BA 6), frontal cortex (BA 9, 10, 44, 45), parietal cortex (BA 7, 39, 40), cingulate cortex (BA 24, 31), and insular cortex (BA 13). Subcortical activations included the putamen, thalamus (ventral anterior nucleus), and claustrum. Quantitative comparison of ALEs across studies of micturition also revealed widespread clusters of activation in both cortex and subcortical brain areas (Figure 2B). Areas of cortex that exceeded threshold included premotor cortex (BA 6), frontal cortex (BA 44), parietal cortex (BA 40), cingulate cortex (BA 24, 32), and insular cortex (BA 13). Subcortical activations included the cerebellum, putamen, thalamus (medial dorsal and ventral lateral nuclei), and claustrum.

To determine whether areas of overlap exist between brain regions activated during swallowing, or when individuals have an urge to swallow, and regions activated when individuals urinate, or have an urge to urinate, we carried out a conjunction analysis between the swallowing and micturition ALE maps. This analysis confirmed that, of the widespread regions activated in both studies, only two areas survived statistical comparison ($p < .05$): the insular cortex of the right hemisphere and the mid-cingulate cortex bilaterally (Figure 3). This finding is consistent with the view, outlined above, that the insular cortex, particularly the right insula, may play an important role in the awareness of bodily sensations, including behavioral urges. It is noteworthy that the two behaviors chosen for study here, swallowing and micturition, involve body areas (i.e., the mouth and oropharyngeal area, and the bladder and genitals) that are located far from one another, and have representations in primary sensorimotor cortex that are also quite distinct and spatially separate.

An fMRI study of the urge to yawn

To further investigate this issue, we sought to investigate the functional anatomy of the urge to yawn. As there are insufficient neuroimaging studies currently published to permit an ALE meta-analysis, we carried out an fMRI study of the urge to yawn. Full methodological information is provided in detail in the online supplementary material.

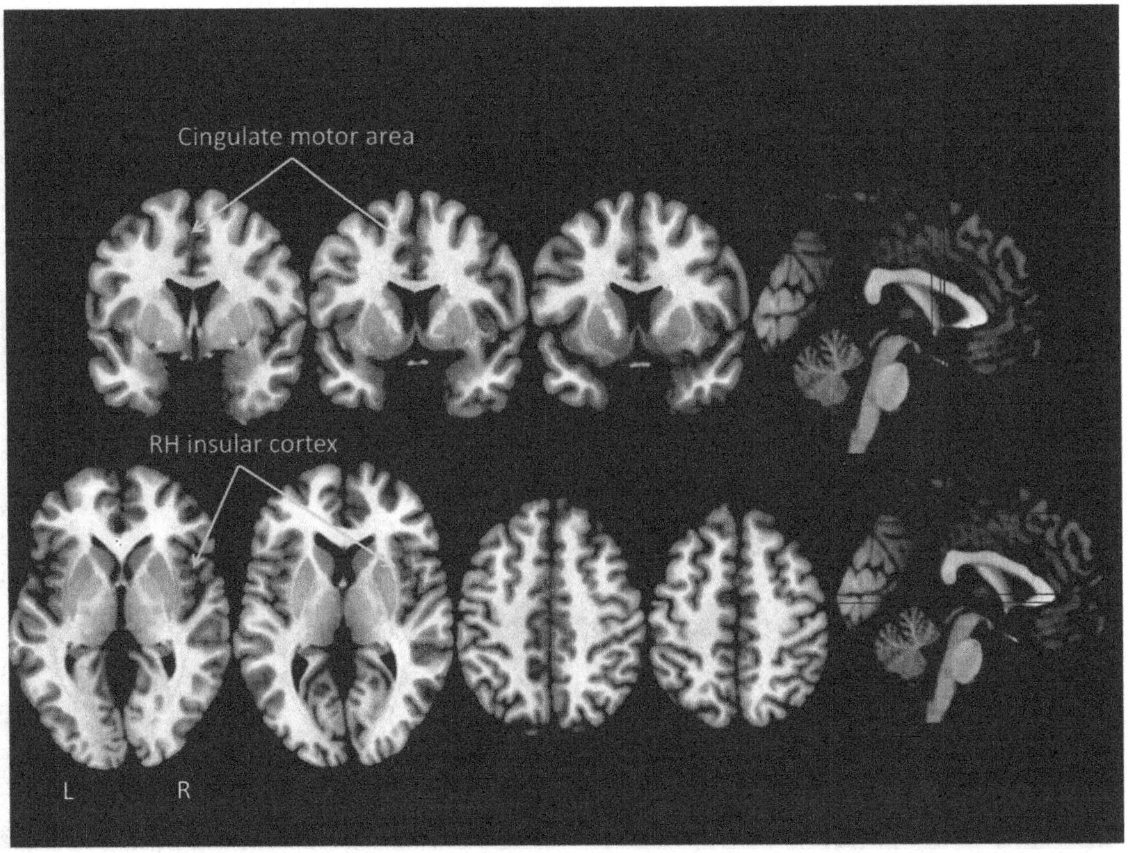

Figure 3. The results of a statistical "conjunction" analysis between the ALE meta-analyses of swallowing and micturition illustrated in Figure 2. The conjunction analysis revealed only two brain areas that reached statistical threshold: the insular cortex of the right hemisphere (RH) and the mid-cingulate cortex (cingulate motor area) bilaterally.

While being scanned, participants observed a sequence of video clips, each approximately 12 s in length, which illustrated an actor yawning. These stimuli had been previously shown to induce high levels of spontaneous yawning in a previous fMRI study (Schüermann et al., 2005). In the current study, participants were instructed to suppress their yawns (which they were able to do successfully) but were required to report periods during which they experienced a strong urge to yawn.

fMRI and anatomic data analysis

Anatomic images were transformed into the Talairach coordinate system and co-registered with each fMRI data set. Regional activation maps were obtained with a single-subject GLM (general linear model) for each individual. We defined a single predictor that modeled the periods that participants reported experiencing the urge to yawn. All 10 individuals included in the analyses reported successfully suppressing their yawns, and inspection of the motion data confirmed this. Second-level analyses involved calculating three-dimensional statistical parametric maps with separate-subject predictors for the group, using a random effects GLM (RFX). The resulting fMRI activity maps were thresholded at a Z value of 3.29 corresponding to $p < .001^{\text{uncorrected}}$ with a minimum cluster threshold of at least 20 voxels. The results of this analysis revealed a number of statistically significant clusters of activation. Details are provided in Table 1.

Details of statistically significant fMRI BOLD activations associated with the urge to yawn are presented in Table 1 and Figure 4. Importantly, the analyses of the fMRI BOLD response associated with participants' subjective reports of their urge to yawn revealed that both the cingulate motor area (CMA) and the insular cortex bilaterally were activated when individuals reported an urge to yawn.

To examine further whether the coordinates of the yawning activations overlapped with the activation foci observed for the ALE analyses of swallowing and micturition reported above, we obtained ALEs for the peak activations of the yawning CMA and insula activations so that they might be compared directly with the results of the

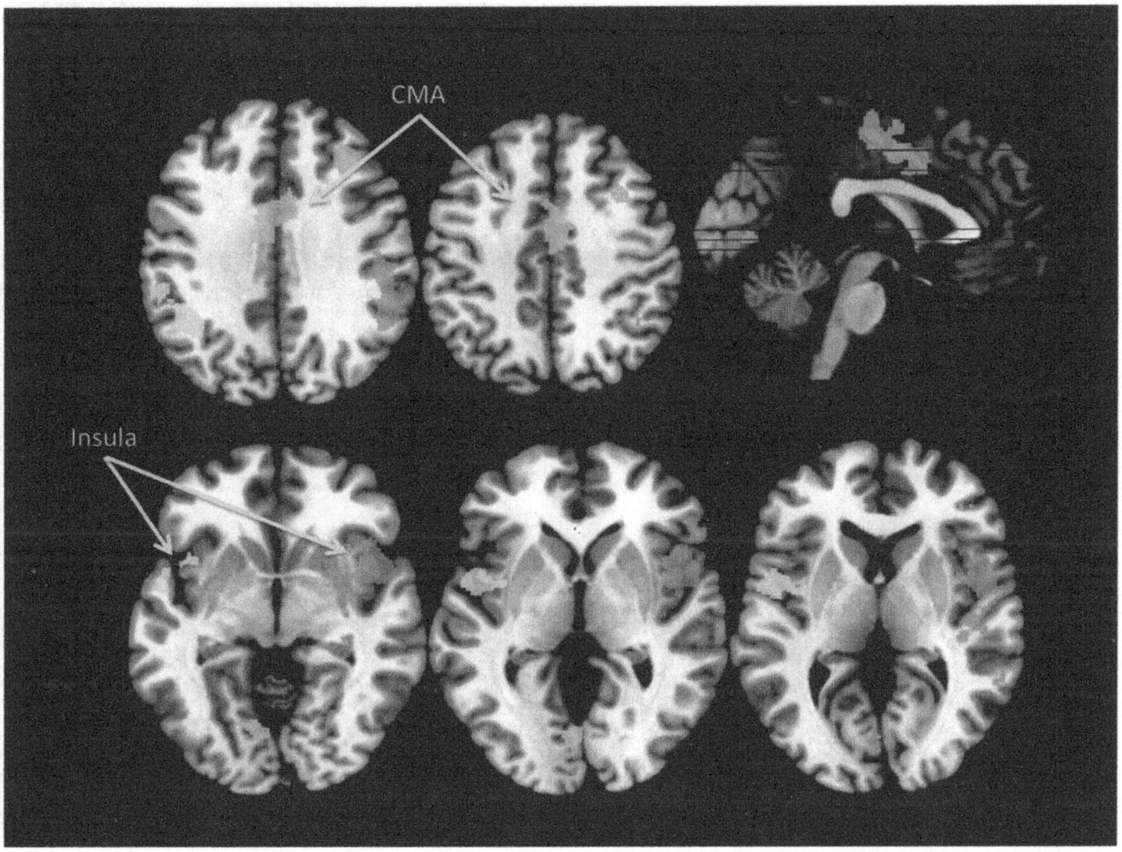

Figure 4. Regions exhibiting a statistically significant increase in blood oxygen level-dependent (BOLD) signal corresponding to the self-reported urge to yawn in an fMRI study of yawning. Again this analysis revealed statistically significant foci of activation within the insular cortex and the dorsal mid-cingulate cortex. CMA: cingulate motor area.

TABLE 1
Coordinates for center-of-gravity and peak activations for statistically significant clusters of activation associated with the urge to yawn

Location	*Mean X*	*Mean Y*	*Mean Z*	*Peak X*	*Peak Y*	*Peak Z*	*Voxels*	*Z value*	*p value*
Right parietal cortex	51.2	−40.8	30.0	50	−32	33	2386	5.85	.00000
Right insular cortex	48.2	4.4	4.9	47	4	−3	3663	6.23	.00000
Cingulate cortex	0.25	−5.0	44.4	−11	−8	30	3361	6.86	.00000
Left parietal cortex	−47.4	−43.5	29.6	−43	−44	27	1713	5.95	.00000
Left insular cortex	−52.3	−7.4	14.5	−43	−20	24	3133	5.89	.00000

ALE results for swallowing and micturition. Furthermore, we also obtained ALEs for the peak activations associated with the urge to tic in individuals with TS reported by Bohlhalter et al. (2006). These data are presented in Figure 5, which illustrates that all four behavioral domains overlap in the region of the mid-cingulate cortex and the insular cortex of the right hemisphere.

Picard and Strick (2001) review the location and functional anatomy of the motor areas located on the medial surface of the human brain. In addition to identifying the supplementary motor area (SMA) and pre-SMA, these authors identify three separable areas within the human cingulate cortex: a caudal cingulate zone (CCZ) situated ventral to the SMA, and a rostral cingulate zone (RCZ) that is further subdivided into anterior and posterior subregions. It is noteworthy that the cingulate fMRI BOLD activations associated with the urge for action in three everyday behaviors (i.e., swallowing, micturition, and yawning), and with the urge to tic in TS, are each located in a region of cingulate cortex that corresponds closely to the CCZ. While the RCZ is associated with conflict detection, attention and arousal processes, and the selection of action, the CCZ is, by contrast, associated with the execution of simple movements and is also activated in response to bodily stimulation such as the delivery of painful cutaneous

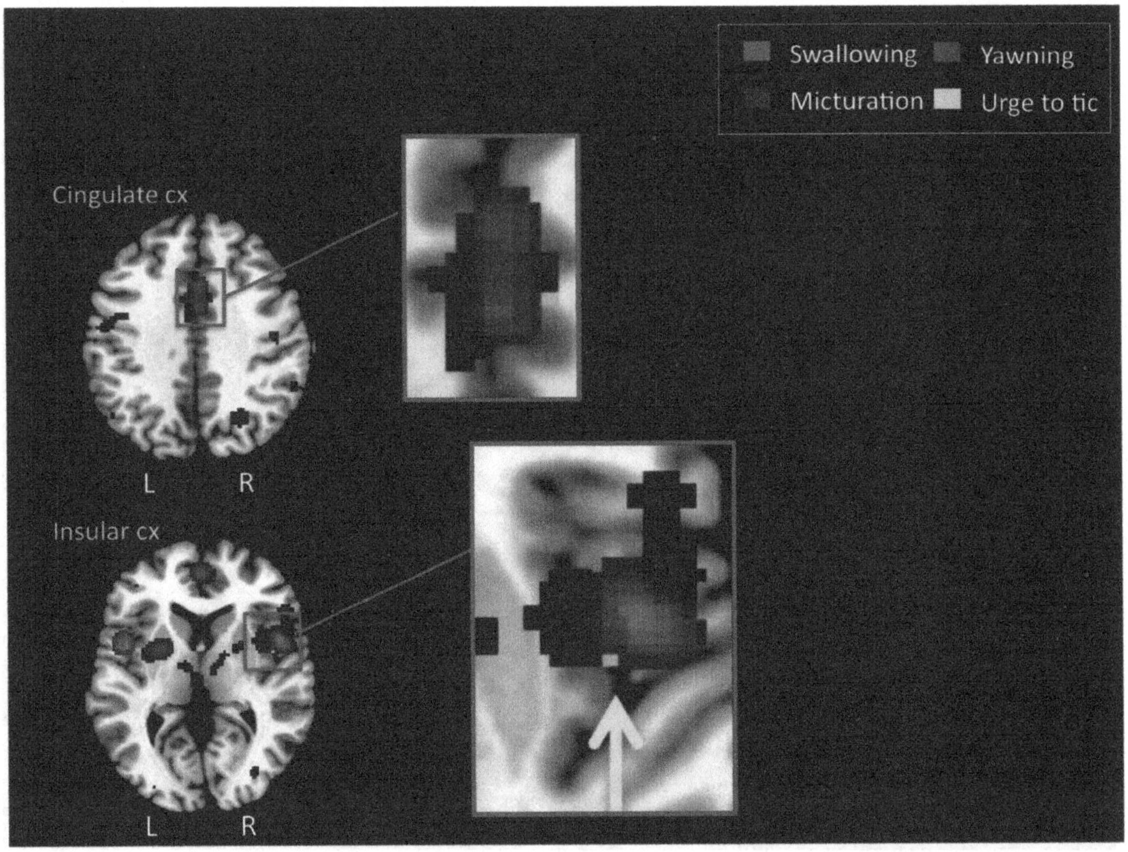

Figure 5. Regions of overlap between ALE meta-analytic studies of swallowing and micturition and fMRI studies of the urge to tic in individuals with TS (Bohlhalter et al., 2006) and the urge to yawn in neurologically normal adults. Again these analyses reveal regions of overlap within the insular cortex (CX) of the right hemisphere and the mid-cingulate cortex bilaterally.

heat and cold stimuli (Picard & Strick, 2001). It should be noted that whereas the SMA has itself been linked to the conscious intention to move (see Desmurget & Sirigu, 2009), and while the CCZ and SMA may be co-activated during movement execution, the SMA and CCA should be considered functionally distinct from one another (Picard & Strick, 2001).

EFFECTIVE CONNECTIVITY BETWEEN INSULAR AND CINGULATE CORTICES

The insula and anterior cingulate cortex have been considered to be the input and output regions of a functional system that is engaged in cognitive, affective, and behavioral contexts (Craig, 2009; Medford & Critchley, 2010). Consistent with this proposal there is now considerable evidence that these regions are jointly active across a wide range of experimental conditions (for a recent review, see Medford & Critchley, 2010). In addition, functional connectivity analyses of resting state fMRI BOLD, using seed regions located within the insula, have shown that (1) the anterior insula is connected functionally with the anterior and mid-cingulate cortex; and (2) the mid- and posterior regions of the insula are connected only with the posterior region of the mid-cingulate cortex (Taylor, Seminowicz, & Davis, 2009).

To investigate the effective, or directional, connectivity of the anterior cingulate and insular cortices, we used the Granger causality mapping (GCM) technique. GCM implements a statistical concept of causality that is based on temporal prediction. Unlike some other methods used to examine effective connectivity, GCM makes no *a priori* assumptions concerning the connectivity of the seed region, but examines the association between the BOLD time course of the seed region and the time course of each voxel outside the seed region. In the current study, two reference or "seed" regions were defined from the BOLD activations associated with the urge-to-yawn versus rest RFX contrast referred to above. The first "seed" region was located within the mid-cingulate cortex, and the second within the insula cortex of the right hemisphere (see the online supplementary material for additional detail).

TABLE 2
Results of GCM analysis of effective connectivity for mid-cingulate cortex and right hemisphere insula "seed" regions

Seed area	*Mean Talairach coordinates*				Center-of-gravity in Talairach coordinates		
	X	*Y*	*Z*		*X*	*Y*	*Z*
Mid-cingulate cortex	0.25	−5.0	44.4				
				Regions exerting an influence over the "seed" area			
				Left anterior insular cortex	−30	20	7
				Right anterior insular cortex	30	20	−2
				Left thalamus: anterior region	−9	−16	−2
				Right thalamus: anterior region	12	−17	1
				Regions influenced by the "seed" area			
				Left mid-insular cortex and inferior frontal lobe	−55	8	2
				Right mid-insular cortex and inferior frontal lobe	52	11	−2
				Left thalamus: centromedial region	−2	−4	7
				Right thalamus: centromedial region	3	−4	6
Seed area	*X*	*Y*	*Z*				
Right insular cortex	48.2	4.4	4.9				
				Regions exerting an influence over the "seed" area			
				Left anterior insular cortex	−42	8	5
				Mid-cingulate cortex	−1	3	44
				Left thalamus: centromedial region	−11	−19	5
				Right thalamus: centromedial region	11	−17	5
				Regions influenced by the "seed" area			
				Left thalamus: ventral lateral region	−14	−14	16
				Right thalamus: ventral lateral region	15	−11	17
				Left thalamus: dorsal medial region	−1	−17	9
				Right thalamus: dorsal medial region	3	−17	8

The GCM analyses revealed a number of brain areas with effective connectivity values that exceeded statistical threshold ($p < .001$). The Talairach coordinates of these areas are presented in Table 2. For the anterior cingulate cortex seed region, the GCM analysis identified four brain areas that likely *exerted an influence over the "seed" region*. These were the anterior portion of the insular cortex in the left and right hemisphere, and the anterior portion of the thalamus bilaterally. These regions are shown in blue in Figure 6. The GCM analysis also identified four brain areas that were themselves *influenced by the "seed" region*. These were the mid/posterior region of the insular cortex in the left and right hemisphere, and the centromedial region of the thalamus bilaterally. These regions are shown in pink in Figure 6.

In addition to revealing patterns of effective connectivity between the thalamus and the cingulate motor areas, the GCM confirmed that the cingulate motor region is influenced by the anterior portion of the insular cortex (i.e., the BOLD response in the anterior insular cortex predicts the BOLD response in the cingulate motor region). This finding is entirely consistent with the view of Craig (2002, 2009) that the insular and cingulate cortices form the input and output regions of a functional system, and can be characterized as the limbic sensory and motor areas. Importantly, the GCM analysis confirms that the cingulate motor region also exerts an influence over the activation of a different region of the insular cortex; specifically the mid-insular cortex. This can be clearly seen in Figure 6A, which illustrates insular regions exerting an influence over the cingulate motor region (blue) and being influenced by the cingulate motor region (pink). Later in this paper, we speculate on the likely functional significance of this anterior insula → cingulate → mid-insular loop.

For the right hemisphere insular cortex seed region, the GCM analysis identified four brain areas that likely *exerted an influence over the "seed" region* (Table 2). These were the anterior portion of the insular cortex in the left hemisphere, the mid-portion of the cingulate cortex (cingulate motor area), and the centromedial portion of the thalamus bilaterally. These regions are shown in blue in Figure 7. The GCM analysis also identified four regions of the thalamus that were themselves *influenced by the "seed" region*. These were the ventral lateral region of the thalamus bilaterally, and the medial dorsal region of the thalamus bilaterally. These regions are shown in pink in Figure 7.

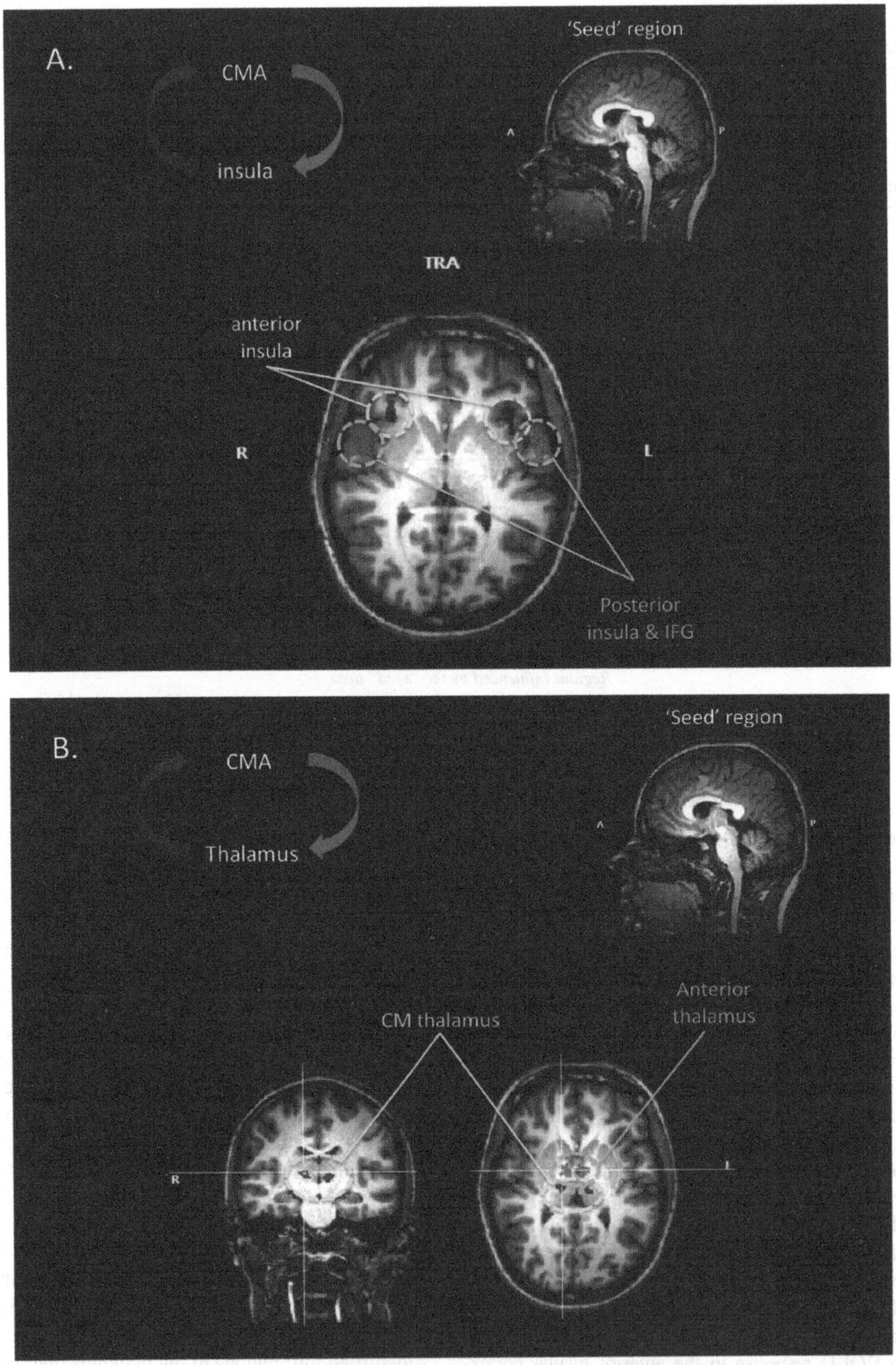

Figure 6. (A) Results of the effective connectivity analyses, based upon Granger causality mapping (GCM), of the yawning fMRI study. In this case, the "seed" region for the GCM has been defined as the region of the mid-cingulate cortex significantly activated during the urge to yawn. The GCM analysis revealed that regions of the anterior insula bilaterally exert a significant influence over the seed region (blue), whereas regions of the mid-insula and inferior frontal lobe bilaterally are influenced by the seed area (pink). TRA: transverse; IFG: inferior frontal gyrus. (B) Further results of the GCM analysis based upon the mid-cingulate "seed" region. The analysis also revealed that regions of the centromedial thalamus bilaterally exert a significant influence over the seed region (blue), and bilateral regions of the anterior thalamus are influenced by the seed area (pink). CMA: cingulate motor area.

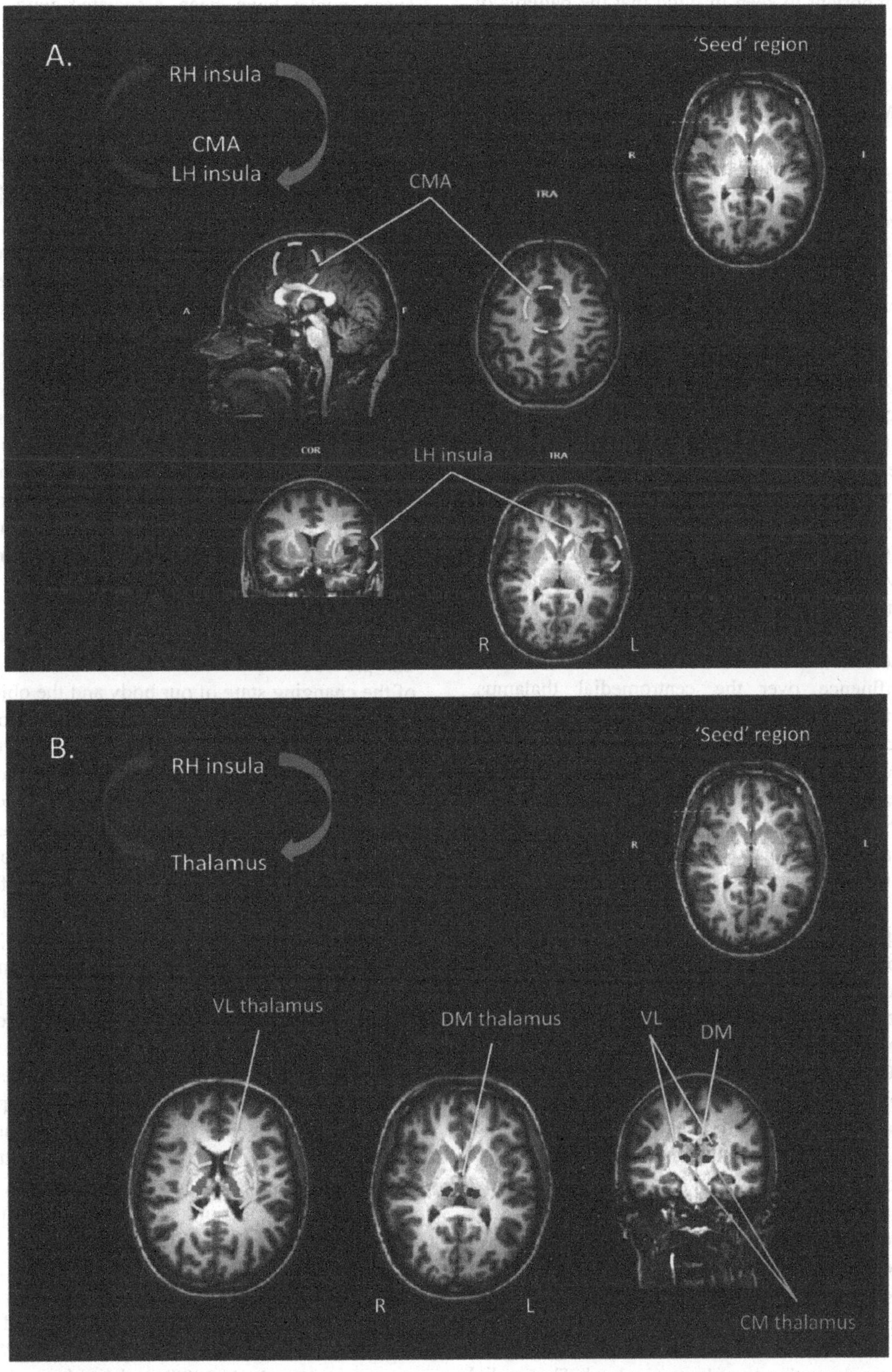

Figure 7. (A) Results of the effective connectivity analyses, based upon Granger causality mapping (GCM), of the yawning fMRI study. In this case, the "seed" region for the GCM has been defined as the region of the insular cortex of the right hemisphere that was significantly activated during the urge to yawn. The GCM analysis revealed that corresponding regions of the insular cortex within the left hemisphere, and the mid-cingulate cortex bilaterally, each exert a significant influence over the seed region (blue). CMA: cingulate motor area; TRA: transverse. (B) Further results of the GCM analysis based upon the right insula "seed" region. The analysis also revealed that regions of the centromedial thalamus bilaterally exert a significant influence over the seed region (blue), and that bilateral regions of the ventral-lateral and dorsomedial thalamus are influenced by the seed area (pink).

It is important to keep in mind that the patterns of effective connectivity revealed by the GCM analyses do not necessarily reflect anatomic connections. Nevertheless, the patterns of connectivity revealed by the GCM are consistent with known anatomy and function of specific thalamocortical connections. For instance, the anterior thalamic nuclei receive their primary input from the hypothalamus, project to the cingulate cortex, and are associated with visceral and emotional processing. In the current study, the GCM analysis confirms that the activity of the anterior thalamus exerts an influence over the cingulate cortex and thus provides a route by which emotional and visceral signals might influence the selection of motor responses.

The centromedian nuclei of the thalamus are a primary source of thalamostriatal projections and play a key role in motor function. In the current study, the GCM analyses confirm that the centromedian region of the thalamus is *influenced* by the activation of the cingulate cortex but in turn exerts an *influence over* the activity of the right insular cortex. This pattern of connectivity suggests that motor processes in the cingulate motor regions of cortex may gain influence over motor selection mechanisms within the striatum via their influence over the centromedial thalamus. Furthermore, the centromedial thalamus may also signal outcomes of the motor selection process to the insula. We speculate that the insular cortex may accumulate evidence on the outcomes of the action, determine whether the conditions giving rise to the urge for action have been resolved, and, if appropriate, generate a sense that the urge for action has been satisfied (see below). In the current context, it is of interest to note that deep-brain stimulation (DBS) of the centromedian region of the thalamus has been demonstrated to be an effective treatment for intractable TS and produces a substantial reduction in the occurrence of tics (Ackermans et al., 2011).

The GCM analysis also revealed that the right hemisphere insular cortex "seed" area is influenced by the BOLD activation within the ventral lateral and dorsal medial regions of the thalamus. The ventral lateral nuclei of the thalamus receive their input from the basal ganglia and cerebellum and send outputs to motor regions of cortex. As illustrated in Figure 1, a key aspect of the motivation-for-action model outlined above is that information about the outcomes of motor events is relayed to the limbic system, where it might be used to determine whether the conditions giving rise to the urge for action have been resolved. The medial dorsal thalamic nuclei are known to receive their inputs primarily from prefrontal and limbic regions of cortex, and project to association areas of the frontal lobe. These nuclei have been associated with complex aspects of cognition, including attentional control and multitasking, goal-directed action planning, and learning and working memory.

A ROLE FOR INSULAR CORTEX IN SENSORIMOTOR PREDICTION?

Our everyday movements often involve interactions between our body and physical objects located within our environment, and engage multiple sensorimotor systems acting in concert. Action selection mechanisms must therefore take account of information about the current state of the motor apparatus (our body) and also of the behavior of objects within our immediate environment. Recent computational neuroscience approaches to the selection and control of movement (e.g., Wolpert, 1997; Wolpert & Ghahramani, 2000) and to reinforcement learning (e.g., Dayan & Niv, 2008) have emphasized the importance of prediction mechanisms to these processes.

It has been argued that efficient motor behavior relies to a large extent upon *predictive* mechanisms that provide accurate estimates, or "internal models," of the changing state of our body and the objects with which we interact. Internal "forward" models are thought to compute dynamic estimates of the body state and to predict the sensory consequences of actions (Wolpert & Ghahramani, 2000). In this view, prediction is necessary to compensate for delays associated with the processing of sensory information, but afferent sensory information is critically important for maintaining the accuracy of internal models and is used to monitor and correct for prediction errors, thereby improving future prediction accuracy and movement control. Thus, any discrepancies between the predicted and observed consequences of an action are used to increase or maintain the accuracy of forward models.

Reinforcement learning also involves the evaluation of the outcomes that follow an action. Computational models of reinforcement learning propose that learning can be based upon internal models of the state transitions and action outcomes within an environment, or based upon model-free learning mechanisms (Dayan & Niv, 2008). Model-free reinforcement learning involves learning to estimate or predict the likely outcome (value or reward) of a given action (state) given an appropriate action-selection policy. This kind of learning is associated with the formation of "habitual" responses (Graybiel, 2008) and has been particularly linked to the operation of the ventral striatum, the neurotransmitter dopamine acting as a reinforcement signal that codes for reward "prediction error" (Schultz,

Dayan, & Montague, 1997). By contrast, model-based reinforcement learning has been associated with goal-directed action in which the efficacy of candidate actions is evaluated with reference to an internal model of the task or state space. A key concept associated with this type of learning is that the selection of an appropriate action may involve a "mental simulation" of potential outcomes (Dayan & Niv, 2008). We suggest that the formation of "urges-for-action" is probably linked to the operation of the habit-learning system, whereas the formation of "desires-for-action" is probably associated with goal-directed, action-planning mechanisms.

A QUANTITATIVE META-ANALYSIS OF REWARD PREDICTION

In this paper, we have speculated that the insular cortex may accumulate evidence on the outcomes of an action and determine whether the conditions giving rise to the urge for action have been resolved. Within the computational framework offered by the forward model literature, this process might involve a comparison of a "next state estimate" (the output of a forward "sensory" model that provides an estimate of the sensory consequences of a planned action) with afferent sensory information that signals the sensory outcomes of the executed action. Within the computational framework offered by model-free reinforcement learning, this process would involve predicting or representing the likely outcome (value or reward) of an action.

To examine whether the insular cortex plays a key role in the representation of the reward estimates for planned actions, we carried out an ALE meta-analysis of functional brain-imaging studies that have investigated the anticipation or expectation of rewards. Full methodological information is provided in detail in the online supplementary material.

Studies were initially chosen by the following keywords: "reward," "prediction," "prediction error," "fMRI" and "prediction," "prediction error," "expectation," and "anticipation." The reward modalities used in the studies reported here were monetary or pleasurable taste rewards. Studies were further refined according to the following criteria.

First, events used in the fMRI contrasts should occur *prior to* the onset of the reward to reflect the outcome of anticipatory processes rather than the processes in response to the actual receipt of a reward.

Second, the contrasts should reflect events in which participants hold an expectation of reward versus a control event such as the expectation of non-rewarding stimuli, or the expectation of loss. The study might include, however, contrasts of high reward probability versus low reward probability. Thus, results that reveal activations by association with the reward probability were explicitly included in the analysis.

Third, participants should expect primary rewards or secondary rewards. Studies that investigated the prediction of non-rewarding events were excluded from this analysis. As a result of these selection criteria, eight studies were included in the meta-analysis. Full details of these studies are provided in the online supplementary material.

The results of the ALE meta-analysis are illustrated in Figure 8. They show that the following brain areas were significantly activated for contrasts associated with the prediction or expectation of rewards across the set of fMRI studies included in the analysis: the left ventral anterior insula cortex, the ventral striatum bilaterally, the globus pallidus bilaterally, and the dorsal cingulate cortex.

The above analyses implicate the input and output nuclei of the basal ganglia together with the anterior insula and the dorsal cingulate cortex in the representation of reward estimates. These findings are consistent with recent proposals that the basal ganglia nuclei modulate movement execution according to motivational factors, specifically context-specific cost/reward estimates (Turner & Desmurget, 2010); that the anterior insular cortex may represent expectations of both positive and negative action outcomes (e.g., Preuschoff, Quartz, & Bossaerts, 2008); and that the anterior cingulate cortex plays a fundamental role in relating actions to their outcomes, and functions, together with the ventral striatum, to mediate cost–benefit decisions over the selection of action (Behrens, Woolrich, Walton, & Rushworth, 2007; Rushworth, Mars, & Summerfield, 2009; Rushworth, Walton, Kennerley, & Bannerman, 2004).

These findings are also consistent with the proposal being advanced here that the anterior insular cortex and the dorsal cingulate motor areas, along with the ventral striatum and thalamus, form core nodes in a motivation-for-action system. Specifically, we propose that pleasant or unpleasant somatosensory or visceral sensations—for instance, the premonitionary sensations that precede the occurrence of tics in TS—are represented within the SII, and in the posterior and mid-insular cortex, and that these sensations will often elicit habitual, overlearned, actions. We also propose that awareness of these bodily sensations, often perceived as an *urge for action*, is associated more specifically with the activity of the anterior insula cortex. Signals associated with such bodily sensations are relayed from the insula to cingulate motor areas that, together with the ventral striatum, may participate

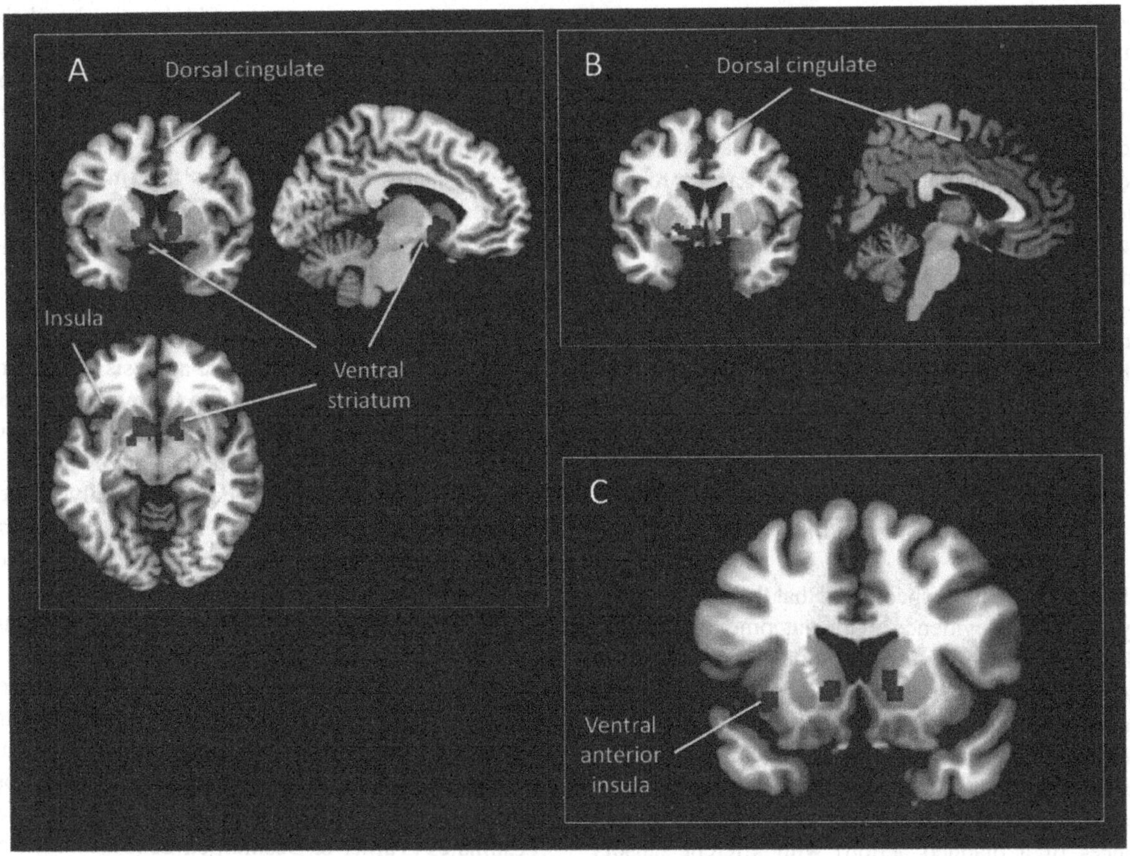

Figure 8. Main results of an ALE meta-analysis of neuroimaging studies of reward expectation (see the online supplementary material for additional information). This analysis revealed a number of activation foci that survived conservative statistical correction ($p < .05$ corrected for false discovery rate). Among these were activations within the insular cortex (A and C) and the dorsal mid-cingulate cortex (B).

in the selection of a particular action based upon a cost–benefit analysis of the likely "value" of an action given the organism's previous history of action outcomes.

Finally, as illustrated in Figure 9, we propose that an estimate of the predicted sensory consequences (outcome) of the selected action is then returned to the insula and inferior frontal cortex, where it may be used to determine whether the sensations giving rise to the urge for action are predicted to alter as a result of the intended action, and, if appropriate, to generate a sense that the urge for action has been satisfied. This last proposal is consistent with recent demonstrations that signals from the cingulate motor area participate in the attenuation of somatosensory activity in SII ahead of voluntary movements (Parkinson et al., 2011). Such sensory cancellation is thought to be a key function of "forward models" (Wolpert, 1997; Wolpert & Ghahramani, 2000), and it is suggested that the predicted sensory consequences of self-generated movements can be attenuated in favor of unpredictable, exafferent, somatosensory input (i.e., sensory signals produced by the environment).

CONCLUSION

In this paper, we have begun to consider the nature and the functional anatomy of "urges-for-action," both in the context of everyday behaviors such as yawning, swallowing, and micturition, and in relation to clinical disorders such as TS, where urges-for-action are considered pathological, and interfere with activities of daily living. We began by reviewing previous frameworks for thinking about behavioral urges, such as Davenport, Sapienza, and Bolser's (2002) motivation-for-action framework that had been formulated in the context of the urge to cough. While it contains several important insights, we took the view that a core aspect of this framework—that actions depend upon the conversion of an urge-for-action into a conscious desire-for-action—was very likely incorrect. We felt that this was particularly true in the case of individuals with TS, in whom urges-for-action were associated with the occurrence of motor and vocal tics that the individuals found both embarrassing and distressing. In this instance, it is difficult to reconcile such actions with the notion that

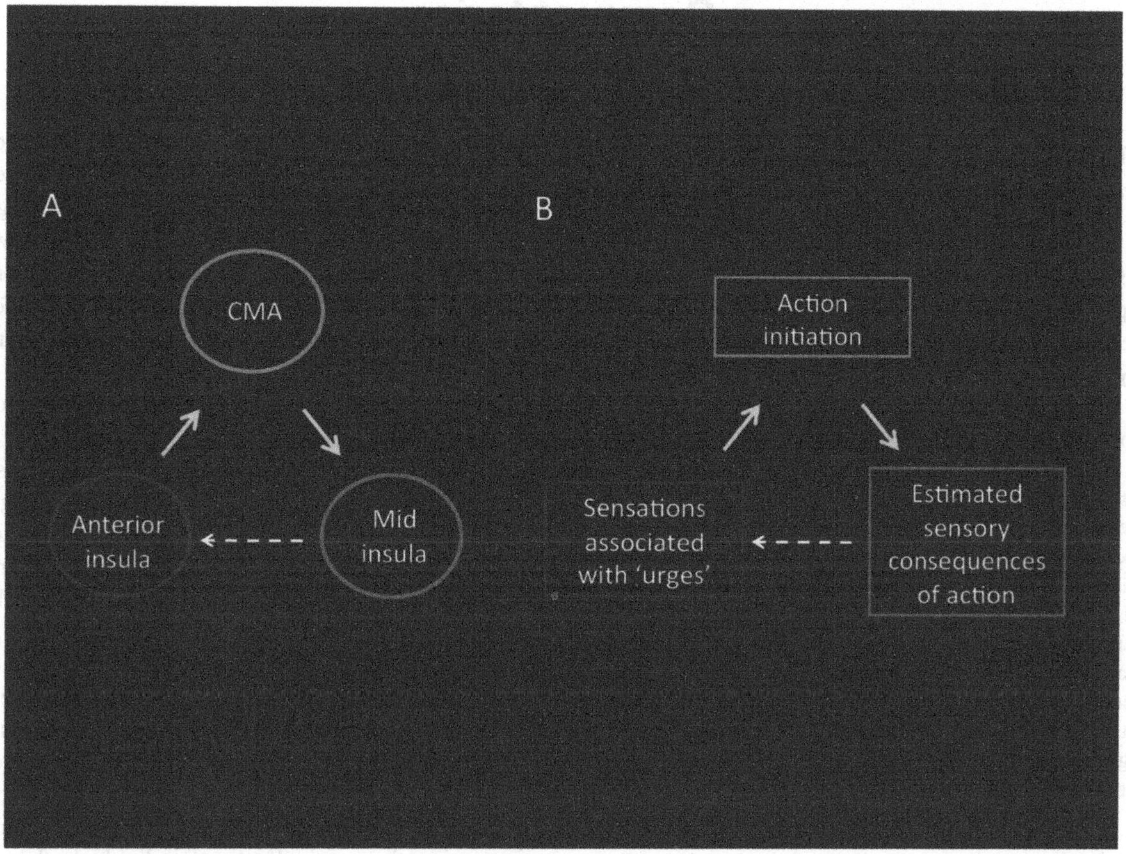

Figure 9. (A) GCM effective connectivity analyses of fMRI BOLD activations of yawning revealed a reciprocal pattern of effective connectivity between the anterior insula, the cingulate motor area, and the mid-insula. (B) On the likely function of these connections, we propose that the anterior insula may represent the urge-for-action, that the cingulate motor region may participate in the selection of an appropriate action following a cost–benefit analysis based upon the organism's past action–outcome history, and that the mid-insular cortex may evaluate, based upon a prediction of the likely outcome of the selected action, whether the conditions giving rise to the urge have been resolved, and, if appropriate, may generate a sense that the urge-for-action has been satisfied. CMA: cingulate motor area.

they spring from a conscious desire for action. Instead we took the view that such actions may be habitual, and as such were highly overlearned and automatic, and could in some cases be executed with very little or no conscious awareness of the sensory stimulation that triggered the action. Importantly, we distinguished between reflex actions and the urge-for-action, which, we argued, occurred when bodily signals gave rise to an action that must be suppressed or deferred.

Then, using quantitative ALE meta-analytic techniques, we investigated the functional anatomy of the urge-for-action in the context of swallowing and micturition, and demonstrated that brain activations associated with these behaviors overlapped in two regions of the brain; the right insula and the dorsal anterior cingulate cortex. Furthermore, we showed that functional activations associated with the urge to tic in individuals with TS and the urge to yawn in neurologically healthy individuals also overlapped within these same two brain areas. These two areas have been conceptualized as the limbic sensory and motor areas respectively (Craig, 2002, 2009), and based upon the effective connectivity analyses of our yawning fMRI data, we proposed that they are central in representing the urge-for-action, and together form a neural circuit that represents bodily sensations, generates an urge-for-action (which may or may not reach awareness), selects a particular action based upon a cost–benefit analysis of the likely "value" of that action, accumulates evidence on the outcomes of that action, determines whether the conditions giving rise to the urge have been resolved, and, if appropriate, generates a sense that the urge has been satisfied. Finally, we argue that this circuit is anatomically separate and largely independent of the neural system responsible for the preparation and execution of intentional, goal-directed, actions, although it is possible that these systems may overlap partially with respect to the sense of agency that accompanies intentional action, which has been associated with the activity of the insular cortex of the right hemisphere.

Commentaries

The urge for self and species preservation

Alice Mado Proverbio
Department of Psychology, University of Milano-Bicocca, Milan, Italy
E-mail: mado.proverbio@unimib.it

http://dx.doi.org/10.1080/17588928.2011.618629

Abstract: Besides sneezing, yawning, and scratching, other types of more complex behaviors can be considered urges for action, in that they are both compelling and can be overtly inhibited. Many behaviors increasing the survival rate of self and species have this instinctual (although not compulsory) nature. We feel a terrible urge to catch a newborn falling from a table, although we could choose not to do so. The hypothesis is advanced that both insula and cingulate cortex participate in social behaviors such as contagious yawning, laughing, and crying, but are not necessarily involved in the urge to cough, for example.

Jackson and coauthors state that motor reflexes (as in recoiling from fire) are different from urge-for-action in that they cannot be cortically inhibited (or temporally postponed). They also characterize urges for actions as different from desires because of their highly implicit and not overly conscious nature. These characterizations are insightful and correct in principle, but the authors focus on very specific and peculiar urge-for-action types, such as the urge of urination, and of yawning and scratching, which are far from homogeneous.

The urge to urinate is one of the many signals that originate in the brain upon hypothalamic impulses and that are involved in the homeostatic regulation of physiological parameters (such as blood oxygen, delta sleep-inducing peptide levels, and plasma concentration of glucose (Burdakov, Luckman, & Verkhratsky, 2005). In this view, the urge to eat in a person with very low glucose levels cannot be considered a reflex (it can be cortically guided, inhibited, or deferred) and is not necessarily a conscious desire (e.g., as with someone who is working and is very concentrated); therefore, it can be legitimately considered an urge-for-action. Similarly, the urge to breathe when there is a lack of oxygen (at high altitude, or in badly air-conditioned places), or, again, the urge to lie down when cerebral blood pressure is low and we feel faint; the urge to quickly undress when it is too hot, or to wrap up when it is freezing; the urge to move arms and legs when we are drowning; and the urge to cough if we are suffocating can all legitimately be considered types of urges-for-action.

We are not sure that insula and mid-cingulate cortex are crucially involved in all these behaviors. It is possible that contagious yawning might share more similarities with other human behaviors reflecting the processing of social information (such as contagious crying and laughing). Indeed, we know that both insula and cingulate cortex play a very important role in empathy, in response to crying (Lang, Yu, Markl, Mü & Kotchoubey, 2011; Sander, Frome, & Scheich, 2007) or to the sight of suffering people (Proverbio, Adorni, Zani, & Trestianu, 2009). Possibly, these structures are responsible for other urges-for-actions such as those regulating instinctual parental behavior (catching a baby falling from a table, for example), aimed at increasing the survival rate of individuals (brushing an insect away from a child) or of the species.

Finally, I would make a distinction between spontaneous yawning and contagious yawning. Spontaneous yawning is thought to have the important function of increasing oxygen blood levels, by expanding the lungs and stretching the muscles (Thompson, 2011). It is also known to lower brain temperature, thus favoring increase in sleepiness and passage to the sleep stage (Gallup & Gallup, 2007). Overall, yawning is probably involved in the circadian and homeostatic control of sleep and wakefulness. Contagious yawning, on the other hand, is thought to be based on the activity of mirror neurons, with the purpose of synchronizing and regulating the behavior of cohabitant individuals, similarly to contagious laughing and crying (Geangu, Benga, Stahl, & Striano, 2010). In this view, the urge to yawn when another does so would stimulate regions involved in theory of mind and empathy, such as the cingulate cortex (Platek, Mohamed, & Gallup, 2005), as was found by Jackson and coauthors.

* * *

Similarities and differences between normal urges and the urge to tic

Beth A. Belluscio, Sule Tinaz, and Mark Hallett
Human Motor Control Section, National Institute of Neurological Disorders and Stroke, Bethesda, MD, USA
E-mail: belluscb@ninds.nih.gov

http://dx.doi.org/10.1080/17588928.2011.618630

Abstract: Investigations into the neurobiological substrates underlying urge are important for developing better understanding and treatment for impulse-control disorders. We characterize the phenomenon based on normal bodily (interoceptive) urges. Features include the following: a preceding awareness of an uncomfortable bodily sensation, a sense of urgency that action must be taken, rising distress when action is delayed, a temporary (e.g., a few minutes in length) ability to suppress or manifest the action voluntarily, subsequent relief once action is taken, association with an action that is necessary to survival. We compare and contrast these characteristics with those described by Tourette syndrome patients as the urge to tic, and highlight several unknowns which merit further investigation.

Understanding the anatomical and physiological basis of "urges" is important not only because they occur in disease states, such as Tourette syndrome, but also because they are closely related to impulses that, when uncontrolled, can wreak havoc on individuals and societies. Problems with impulse control include substance abuse, domestic violence, sexual deviancy, and pathological gambling. In this context, it is important to clearly define what is meant by "urge," to delineate what distinguishes normal from pathological urges, and to clarify how urges differ from other modes of action selection. Study of the underlying functional anatomy, as reviewed by Jackson et al., can then lead to a greater understanding and better treatment of impulse-control disorders.

The elementary components of brain function can be divided into sensory input, processing, and motor output. Urge is one mode of processing. Although it exists within a continuum of modes including reflex, desire, and intention, in which boundaries are somewhat artificial, defining the phenomenon helps unify research efforts. Sensory input that leads to urge can come from within the body (interoceptive) or from outside the body (exteroceptive). We characterize urge based on examples of common bodily urges: to urinate, yawn, cough, blink, and sleep. In each, sensory input can trigger the action out of awareness, in which case there is no urge and the action is perceived as occurring involuntarily. If, however, the action is delayed, the sensation reaches awareness and the urge to act develops. Urge is separate from the sensation. The sensation that our bladder is full is distinct from the *need* (urge) to empty it. Bodily sensations that trigger urge are uncomfortable and, as they mount in intensity, lead to a feeling of distress. This is probably because the actions to which urges lead are those that are essential to our survival. During a brief period, we can control the action by either suppressing it or executing it, but at some point it occurs without our will. In this way, the actions driven by urges are neither truly involuntary nor truly voluntary. Because of the short time frame in which action must be taken, there is a sense of urgency with urge, differentiating it from desire. Once the action is taken, the sensation is eliminated and there is an experience of relief.

In many ways, Tourette patients experience a parallel series of events (Bliss, 1980). Their tics can occur completely out of awareness, and, in these instances, would be described as involuntary. Often, however, patients describe an uncomfortable bodily sensation that triggers the urge to tic: a feeling that the action *must* be done. When tics are suppressed, the sensation and urge increase in intensity, leading to mounting distress. Patients state that they voluntarily make tics occur in response to the urge (Lang, 1990), causing a reduction in the bothersome sensation and a feeling of relief. With prolonged suppression, the urge becomes overwhelming and tics occur beyond patients' control. Thus, tics reside in that gray zone between involuntary and voluntary, and have been called "un-voluntary."

However, there are some key distinctions between normal interoception-related urges and those associated with tics. First, we do not yet know what type of sensory input generates the uncomfortable bodily sensations that trigger the urge to tic. Although patients sometimes describe these sensations as coming from muscles or joints, it has never been demonstrated experimentally that sensory activity initiated in the periphery is the origin of their sensory experience. Second, tics are not key to survival. It is an important aspect of the pathology in Tourette syndrome that patients experience such distress in relation to actions that are not only nonessential but also nonproductive. We do not yet understand when or how this association is made. Third, following the tic, the bothersome sensation is not eliminated, but only reduced in intensity. Finally it is not clear why tics must be performed in a particular way in order to achieve relief and diminishment of the bothersome sensations. Continued research into the circuitry that mediates both normal and pathological urges, as described by Jackson et al., will help our understanding and treatment of Tourette syndrome and other impulse-control disorders.

Features of normal and tic-related urges

Characteristic	*Normal*	*Tic*
Sensory input from bodily organs	Yes	?
Sensation reaches awareness when action delayed	Yes	Yes
Sensation is uncomfortable	Yes	Yes
Experience urgency and need to take action	Yes	Yes
Action can be suppressed (minutes), and then must occur	Yes	Yes
Relief from discomfort with action taken	Yes	Some
Action is necessary to organism survival	Yes	No

* * *

The role of consciousness in the urge-for-action

Álvaro A. Rivera-Rei[1], Andrés Canales-Johnson[1], David Huepe[1], and Agustín Ibáñez[1,2]
[1]Laboratory of Cognitive and Social Neuroscience, Universidad Diego Portales, Santiago, Chile
[2]Institute of Cognitive Neurology (INECO), Favaloro University, and National Scientific and Technical Research Council (CONICET), Buenos Aires, Argentina
E-mail: aibanez@neurologiacognitiva.org

http://dx.doi.org/10.1080/17588928.2011.618631

Abstract: A neuroanatomical model of urge-for-action phenomena has been proposed based on the "motivation-for-action" network (e.g., insula and mid-cingulate cortex). Notwithstanding the sound evidence presented regarding the functional and anatomical correlates of this model, the nature of the relationship between urges and conscious awareness remains to be addressed. Moreover, this model does not seem to explain (1) how a conscious access threshold is reached, and (2) the way in which the urges are related to more general contents of consciousness.

Jackson et al. have proposed a novel model of urge-for-action. This model considers the nature and functional anatomy of urge-for-action in the context of normal life and clinical disorders. Through a meta-analysis, the authors show that there is an overlap between limbic sensory and motor neural circuits related to urges of everyday behaviors such as swallowing and tics in Tourette syndrome. The primary merit of this work is unquestionable; the authors propose an empirical and theoretical model of urge-for-action incorporating actions that do not necessarily require conscious awareness of the sensory stimulation that triggered them. Nevertheless, there are important issues that this model does not explicitly incorporate.

First, it is not clear how urge-for-action, as defined in this paper, can be a fully unconscious phenomenon. If the momentary inhibition of the action is an integral part of the definition of these urges, then that inhibition should be unconscious as well. In this paper, we find a lack of evidence on this topic, perhaps reflecting more than a mere lack of interest on this matter. Having searched the literature on unconscious inhibition of action, we found several instances in which willed intention is involved (e.g., go/no go tests in Eimer & Schlaghecken, 2002, and van Gaal, Ridderinkhof, van den Wildenberg, & Lamme, 2009). It is hard to imagine a fully unconscious inhibition of the kinds of actions considered to be representative of urge-for-action (coughing, swallowing, yawning, etc.). As long as that inhibition plays an important role in this process, we should be consciously aware of our urge-for-action as opposed to our reflexes.

The second issue concerns the nature of the relationship between urges and conscious awareness. Jackson et al. explained that the intensity of physiological afferent stimulation relates directly to the awareness of the urge's strength during a phenomenon such as swallowing, but it is not clear how this phenomenon could be explained by their proposed model.

Despite the straightforward relationship between insular cortex and interoceptive conscious awareness (e.g., Ibáñez, Gleichgerrcht, & Manes, 2010), this fact is not explicitly taken into account in terms of the neuronal activity of the cortical and subcortical regions considered in this model. More importantly, even if we consider that urge-for-action could be an unconscious phenomenon, it remains unclear how this model could explain the transition between urges the subject is not conscious of and those of which the subject is consciously aware. Along these lines, is it the strength of activation of the right insular cortex, the anterior cingulate cortex (ACC), or the circuit between these regions that is responsible for setting the threshold between unconscious and conscious awareness of urges? This point is far from being addressed in Jackson et al.'s model.

Finally, how could the urge-for-action model of Jackson et al. be integrated with more general models of consciousness? Current models have determined the activation of widespread cortical regions during goal-directed visual awareness (e.g., Dehaene et al., 2001) and have specified some neuronal markers for reaching the threshold of conscious perception (e.g., Del Cul, Baillet, & Dehaene, 2007). How does this urge-for-action circuit interact with more general circuits of conscious perception? Interoceptive awareness is conceptualized as the capacity of being aware of some specific *content* of consciousness; that is, of visceral perception (Ibáñez et al., 2010). Therefore, the authors should clarify the way in which interoceptive awareness might be related to other contents of consciousness, such as goal-directed cortical circuits underlying more general conscious perception phenomena.

* * *

Urges, inhibition, and voluntary action

Parashkev Nachev
Institute of Neurology, University College London, London, UK
E-mail: p.nachev@ion.ucl.ac.uk

http://dx.doi.org/10.1080/17588928.2011.618632

Abstract: It is constitutive of the notion of an urge that it must precede the action it urges. For the duration of an urge to be non-zero, some process must keep the action being urged in check. Urges therefore inevitably involve inhibition of action, and perhaps conflict between action and inaction. In any event, they cannot form a critical part of the phenomenology that many argue must precede voluntary action, for if they play any part at all, it is only in situations where the action is to some degree inhibited.

The notion of "urge" has come to play a pivotal role in current conceptions of voluntary action. Curiously, it has done so without detailed exploration of the kinds of actions—involuntary or atypically voluntary—where the need to invoke it is arguably much greater, and where an understanding of its neural basis is far more plausibly within reach. And when one analyzes such actions—as Jackson and his colleagues lucidly and comprehensively do—the notion of urge that emerges is radically at odds with one that gives it a critical role in voluntary action. This is just as important a conclusion as the independent clarification of the neural substrate of urges itself.

It is surely right that urges must be dissociated from the sensations that in some cases prompt them. The urge to yawn, for example, is not plausibly any kind of sensation: One can only describe it with reference to the action it compels. What is minimally constitutive of the notion of an urge is that it is an urge *to do something*: it is both transitive and directed at a specific action. It is also true, as Jackson and his colleagues point out, that an urge must temporally *precede* the action: If it post-cedes it, we would not call it an urge, and if it parallels it in time, then it could not be an urge to perform the action because the action is already being performed. But the implication goes further. Since the urge is to perform the action, the necessary interval between it and the action during which the urge is experienced must involve *inhibition* of the action, for otherwise the action would be performed immediately. Inhibition is therefore an inevitable consequence of the notion of an urge.

Indeed, it is striking that the actions commonly associated with urges—both in the normal and in the pathological state—are usually of the kind that cannot be actively *chosen* but only *withheld*. One cannot actively choose to yawn or sneeze, and if one coughs or voids one's bladder voluntarily one tends to do so pre-emptively of the normal action. To the extent to which we have control over such actions, it is to keep them in check, to be released at the time when their performance is convenient. There are, of course, other comparably "automatic" actions, such as blinking, where urges play a much less prominent role, but it is notable that they tend to be of a kind one rarely has reason to suppress. Equally, the actions associated with urges in movement disorders such as Gilles de la Tourette syndrome are perceived as unwanted; where no such strong perception exists—in many cases of chorea, for example—patients do not complain of urges.

An urge to act, then, cannot be empirically dissociated from the inhibition of the action, and presumably also from the conflict between action and inaction: this is so not because we lack the experimental tools but because inhibition is constitutive of the notion of an urge. This implies that the neural substrate Jackson and his colleague identify must also subserve these processes. More importantly, it casts further doubt on the notion that an urge is a signature of the phenomenology that is argued to precede voluntary action and to take part in its self-ascription. For if an urge signifies the inhibition of an action, it can hardly be thought of as the "idea" driving it. This also explains why urges are often reported during macrostimulation of the dorsomedial frontal cortex: The set of multiple neuronal pools inevitably severally activated at that scale of stimulation is bound to contain units that inhibit as well as drive active movement.

There are, of course, numerous unanswerable arguments against the ideomotor theories of action popularized by Libet's work, but keeping the lid on that coffin is so extraordinarily difficult another nail can never go amiss.

* * *

Unaware urges? Let's not complicate matters further

Edward H. F. de Haan
Department of Psychology and Centre of Cognitive Sciences Amsterdam, University of Amsterdam, Amsterdam, The Netherlands
E-mail: e.h.f.dehaan@uva.nl

http://dx.doi.org/10.1080/17588928.2011.618633

Abstract: The model put forward for the neuroanatomical basis of urges-for-action is compelling. The arguments based on meta-analyses of existing neuroimaging data are elegant and convincing. However, I am not convinced by the suggestion that there are conscious urges and urges that remain unaware. In my view, awareness is a defining feature of an urge.

Jackson et al. develop a convincing case for the existence of a neuroanatomical system that is responsible for the "urge-for-action." The role of the (right) insula and the dorsal anterior cingulate cortex is inferred on the basis of a number of separate meta-analyses looking at associated brain activations in response to specific stimulation—in this case different triggers of urges. The idea that there are separate systems for "willed" or planned actions and "urged" actions makes sense and is clearly supported by the clinical and neuroimaging data.

The definition of "urges-for-action" proves more complicated. The beginning is relatively straightforward. These urges involve a limited set of actions that are habitual, standardized routines, such as scratching, yawning, or swallowing. They typically do not require feedback, and they are functional in that they alleviate specific negative or unpleasant bodily states. Things become tricky when we are offered the suggestion that we have conscious and unconscious urges (cf. Dijkerman & De Haan, 2007). We are told that urges and desires are not synonymous, and the main reason for this is that urges may also remain unaware to the person who "experiences" them. Indeed, we may find ourselves swallowing or yawning without having had a conscious urge, but it is not clear to me why this behavior cannot be classified as "reflexive." Indeed, the Davenport, Sapienza, and Bolser (2002) study seems to indicate that we are dealing with a system where the intensity of the (negative) stimulation is directly related to strength of the perceived urge. This makes sense, the limbic system including the insula, registers the amount of bodily discomfort, and depending on the severity of this discomfort, a more or less "urgent" urge is felt.

I also agree with the authors that what distinguishes a reflex from an urge is the fact that a reflex—by definition—is immediate and proceeds without conscious interference. An urge, however, may need to be postponed as the execution of the required action is socially or emotionally compromising. Here I would like to suggest a different interpretation from the one put forward by Jackson et al. that an urge may be construed as an interrupted reflex. The reflex entails a complex sequence of detecting a specific negative bodily state coupled to a proven alleviating motor action. The fact that this motor action is not socially or emotionally acceptable is in essence culturally defined, and therefore, is a learned response.

Thus, and this is my central thesis, the defining aspect of an urge is a learned interruption of a reflexive sequence. The result of this interruption is a conscious awareness of the pending action *and* the fact that this action is not acceptable in the current situation. In this view, an unconscious urge is a contradiction in terms, as an urge constitutes the awareness of the desire to perform an action coupled with the learned response to inhibit this action.

Jackson et al. state that it is "difficult to reconcile the concept of a 'desire,' which is often defined as 'a longing or craving for something that brings satisfaction or enjoyment,' with unwanted actions, the execution of which is experienced as unpleasant and distressing" (p. 229). I find this surprising and I cannot subscribe to this "difficulty" felt by the authors, as in my experience this is often exactly what patients with obsessive compulsive disorder or Gilles de la Tourette syndrome report. The simultaneous experience of the urge to carry out an action and the concurrent realization that this is not acceptable, socially or emotionally, could be posited as the hallmark of obsessive compulsive disorder. In fact, after the act has been performed, there is again a simultaneous experience but now one of relief (because the act has—somewhat—alleviated the negative bodily state) and shame (because the executed act was socially or emotionally unacceptable). Interestingly, when I was reading the article for the first time, I thought that, toward the end, the authors would propose a role for the right insula in becoming aware of an urge. They did not do so, and I am glad they did not, because it is not important for their main conclusion.

This review provides strong evidence for a separate network in the brain involved in the detection of well-defined negative bodily states and a set of habitual, standardized routines to alleviate the discomfort. When the required action is consciously flagged because of learned associations with negative social or emotional connotations, the execution of these actions is— temporarily—inhibited, and an "urge" is experienced.

* * *

My urge, my tic – a missing link between urges and tic inhibition

C. Ganos[1] and F. C. Hummel[2]
[1]Movement Disorders Research Group, Department of Neurology, University Medical Center Hamburg-Eppendorf (UKE), Hamburg, Germany
[2]Brain Imaging and Neurostimulation (BINS) Laboratory, Department of Neurology, University Medical Center Hamburg-Eppendorf (UKE), Hamburg, Germany
E-mail: f.hummel@uke.de

http://dx.doi.org/10.1080/17588928.2011.618634

Abstract: Despite the fact that premonitory urges precede most tics in patients with Gilles de la Tourette syndrome (GTS), the voluntariness of tic elicitation and its suppressibility as a response to these urges still remains unclear. Moreover, there are no systematic studies examining the association between urge intensity and the ability to suppress tics. As shown by behavioral, neurophysiological, and imaging data, sensorimotor networks in GTS exhibit altered patterns of organization modulated through interactions with frontomesial networks of volitional inhibition.

In their nice and elaborated review, Jackson et al. provide a functional basis for a distinction between reflexive behaviors and actions that come from urges helpful in daily life (such as the urge to yawn, urinate, and cough) and those possibly interfering with daily life (such as the urge to tic). They argue that the latter are associated with *unwanted actions*, principally as their *interoceptive* gatekeepers, allowing their conscious *suppression or deferment*. Particularly for Gilles de la Tourette syndrome (GTS), they propose that premonitory sensations are represented within the somatosensory (SII) and mid- and posterior insular cortices, leading to *habitual actions*. They argue that the urge for action is associated with activity of the anterior insular cortex, which is then relayed to the cingulum and the ventral striatum for reward-based prediction analysis, and then forwarded again to the insular and inferior frontal cortices, which in their turn satisfy or propagate the urge for action.

Examining the properties of the urge in GTS, and already before Leckmann's seminal paper (Leckmann,

Walker, & Cohen, 1993) on the subject, Bliss, a clinician and GTS patient himself, published (Bliss, 1980) a thorough description of his own premonitory sensations and, as a result, argued for the voluntariness of tics. In the following years, systematic studies showed that more than 90% of patients with GTS report these urges and consider tics as, if at all, only partial involuntary. It has been consistently found that the development of the urge is reported with a lag of about 3 years after the onset of tics (Banashewski, Woerner, & Rothenberger, 2003; Kwak, Dat Vuong, & Jankovic, 2003; Leckmann et al., 1993). The delayed appearance of this phenomenon has led to a series of hypotheses: Is the association of urges and tics a compensatory evolutionary development as a basis for the option to suppress these phenomena, or are urges and tic suppressibility co-existing phenomena, as suggested by recent tic-suppression studies? (Banashewski et al., 2003; Conelea & Woods, 2008).

As for the neurophysiology of tic generation, the zeitgeist led to the study of the correlates of the voluntariness of movement disorders, with the pioneering work of Obeso, Rothwell, and Marsden (1981), followed 14 years later by another work (Karp, Porter, Toro, & Hallett, 1996), which examined the presence of the Bereitschaftspotential (BP) in tics. The results were equivocal, with the first study showing that tics were not preceded by the BP, and the second demonstrating its presence in 2 out of 5 patients. This led to the hypothesis (Kwak et al., 2003) that only tics associated with a premonitory sensation—and therefore consciously perceived and voluntarily initiated—are preceded by the BP. This has been shown in three patients in the only study to date that addressed this question (Duggal & Nizamie, 2002). Additionally, and in accordance with the latter study, which, surprisingly, found a shorter than normal BP, Moretto showed that patients with GTS have a delayed experience of volition (Moretto, Schwingenschuh, Katschnig, Bhatia, & Haggard, 2011). This would imply that not only the formation of tics but also the formation of normal movements would necessitate an altered pattern of motor organization through fronto-striato-thalamo-cortical pathways, as supported by current findings (Heise et al., 2010; Roessner et al., 2011). Furthermore, in the absence of direct evidence to support the notion that the awareness of urges correlates positively with the ability to suppress tics—a suppression, which, according to Jackson et al., would lead to the propagation of the urge and further activation of the anterior insula—the role of frontomesial networks of volitional inhibition (Kühn, Haggard, & Brass, 2009) and tic suppression has to be addressed in detail. These networks presumably influence the pre-supplementary motor area activity in a top-down fashion, as supported by a recent EEG experiment showing elevated inter-regional interactions between these and sensorimotor and prefrontal areas during tic inhibition, paralleled by an increasing urge to tic (Serrien, Orth, Evans, Lees, & Brown, 2005).

* * *

An urge to act or an urge to suppress?

John C. Rothwell and Mark J. Edwards
Institute of Neurology, and Sobell Department of Motor Neuroscience and Movement Disorders, University College London, London, UK
E-mail: j.rothwell@ion.ucl.ac.uk

http://dx.doi.org/10.1080/17588928.2011.618635

Abstract: There is no doubt that there exist urges to act, but are these really the cause of the action or is an urge a shorthand term to describe the effort involved in suppressing an action that is triggered automatically by some other process?

There is no doubt that there exist urges to action— just think back to Zidane's uncontrolled and badly timed urge to head-butt an opponent. But can we go quite as far as Jackson and colleagues in believing that they all arise from activity in the same anterior insular and caudal cingulate regions of the cortex? Furthermore, does activity in these regions then drive the premonitory sensations that are sometimes described by patients with Tourette's syndrome prior to their tics? The argument of Jackson and colleagues is erudite and informed, but there seems to be a certain sleight of hand in the logical steps of the argument which lead to the definition of a "motivation for action" network.

The authors begin with the highly reasonable assertion that an urge to act is, at least in the majority of instances, an awareness of the effort involved in restraining the act. That is, some other factor (perhaps only dimly perceived) is the trigger for this latent action, and when we become aware of the impending action, we are able to prevent it from being released by effort of will. Effectively, this seems to be saying that stimulus–response associations exist that can give rise to actions. Consistent with this idea, the authors point out that some actions that can be associated with urges on one occasion (e.g., yawning) can, on another occasion, emerge without the sensation of an urge. The rule seems to be that if the stimulus or impending action is perceived before the action occurs, and if the action is subsequently withheld, then we experience an urge to act (unless the stimulus disappears). We can feel an itch or a "desire" to scratch and we can decide to withhold the scratch; if the itch persists, then we may develop an urge to scratch. Paradoxically, an urge to act turns out to be an intention to suppress. The action itself is triggered by some other factor in an automatic fashion.

This seems to be a very simple and reasonable interpretation that involves two interacting processes: the basic stimulus–response coupling and a supervisory system with a power to withhold the response. However, the authors then seem to pursue quite a different interpretation about halfway through the article. The turning point comes after the analysis of the first set of imaging data. Here they find common activation in anterior insula and caudal cingulate cortex during yawning and micturition. They then say that activity in these common areas is responsible for the urge to act and then causes the action, as illustrated in Figure 9. At this stage, we have lost the low-level stimulus–response coupling and all mention of inhibition. We are now told that inputs to the anterior insula produce an urge to act (not to suppress a prepotent action). This constitutes a feeling that we have to make a particular movement, which is then achieved via activation of caudal cingulate cortex.

We accept that it is difficult to decide which of these interpretations is true. What strikes us as odd is that the authors begin by emphasizing the importance of inhibition, yet end by talking about motivation to act. Would it not be simpler to suppose that stimulus–response associations exist at all levels of the sensorimotor system from spinal reflexes to striatal habits. These are all to a greater or lesser extent modifiable by a supervisory system we may equate with volitional control. An urge to act is an expression of the interaction between these systems, not a separate system itself.

* * *

 http://dx.doi.org/10.1080/17588928.2011.618628

Reply to Commentaries

Resolving confusions about urges and intentions

Stephen R. Jackson[1,2], Amy Parkinson[2], So Young Kim[1], Martin Schüermann[2], and Simon B. Eickhoff[3,4]

[1]Department of Brain and Cognitive Engineering, Korea University, Seoul, South Korea
[2]School of Psychology, University of Nottingham, Nottingham, UK
[3]Department of Psychiatry and Psychotherapy, RWTH Aachen University, Aachen, Germany
[4]Institute of Neuroscience and Medicine (INM – 2), Research Center Jülich, Jülich, Germany

In our article we considered the nature and the functional anatomy of "urges-for-action," both in the context of everyday behaviors such as yawning, swallowing, and micturition, and in relation to clinical disorders in which the urge-for-action is considered pathological (e.g., Tourette syndrome), and we argued for a key role for the insular and cingulate cortices in experiencing the urge-for-action. Here we seek to address some of the key points raised within several of the interesting commentaries on, and challenges to, our paper.

Our intention when writing our article was to instigate discussion on the nature and functional anatomy of the urge-for-action, particularly with reference to the urge to tic in Tourette syndrome. We are therefore most grateful to the many authors who engaged with this issue and contributed commentaries on our paper. The issues raised within the commentaries are useful and insightful, add greatly to the discussion of this topic, and will no doubt be of interest to the readers of this journal.

The commentaries focused on several core issues with respect to the framework for thinking about the urge-for-action introduced within our paper. These included the distinction between reflexes, urges, desires, and intentions, and the role played by consciousness in the urge-for-action (e.g., Belluscio et al.; de Haan; Rivera-Rei et al.; Nachev); the part played by inhibition in defining urges-for-action (e.g., Rivera-Rei et al.; Nachev; Rothwell & Edwards); the involuntary or (un)voluntary nature of urge-related behaviors in disorders such Tourette syndrome or obsessive-compulsive disorder (OCD) (e.g., Belluscio et al.; de Haan; Ganos & Hummel); and the role played by the insular and cingulate cortices in experiencing urges-for-action (e.g., Proverbio; Rothwell & Edwards) Below we briefly respond to each of these issues.

DISTINCTION BETWEEN REFLEXES, URGES, DESIRES, AND INTENTIONS, AND THE ROLE PLAYED BY CONSCIOUSNESS IN THE URGE-FOR-ACTION

Both de Haan and Rivera-Rei et al. argue against the idea that urges can be unconscious, and de Haan proposes that bodily sensations that trigger actions outside awareness should be labeled reflexes. We agree entirely with this position, and it was never our intention to suggest that urges could occur outside awareness. In fact, we argued for precisely this distinction when we proposed that "One possible distinguishing feature of urges, as distinct from reflexes, may be that urges are chiefly associated with actions that cannot be realized immediately and must be held in check until an appropriate time when they might be released."

However, we agree very much with the comment made by Belluscio and colleagues that the boundaries between the concepts of reflex, urge, and intention are rather artificial, and with their analysis that "sensory input can trigger the action out of awareness, in which case there is no urge and the action is perceived as occurring involuntarily. If, however, the action is delayed, the sensation reaches awareness and the urge to act develops." Nevertheless this does raise an interesting question concerning the functional role played by awareness. Specifically, does awareness in any way alter the nature of the action performed? Belluscio and colleagues appear to suggest that whether a sensation reaches awareness is a matter of degree, and imply that the action that follows a sensation is the same irrespective of awareness. Thus, we might ask: is the yawn that occurs spontaneously, without awareness of an urge to yawn, in any way different from the yawn that occurs in spite of our attempts to stifle it? This position seems reasonable and consistent with the observation that actions can be triggered by sensory stimuli without awareness, and with previous findings that the intensity of sensory stimulation predicts the strength of the urge that is experienced (Davenport, Sapienza, & Bolser, 2002). By contrast, Rivera-Rei and colleagues appear to suggest that there may be something important to be explained with respect to the "transition between urges the subject is not conscious of and those of which the subject is consciously aware." We suggest that the ambiguity over the concepts of reflex, urge, and intention may be resolved when we have a better understanding of the role that consciousness plays in the initiation of action as advocated by Rivera-Rei and colleagues.

Several commentators either explicitly (Belluscio et al.; Nachev) or implicitly (Rothwell & Edwards) argue for a distinction between our experience of a bodily sensation and our awareness of an urge for action. Belluscio et al. argue, "Urge is separate from the sensation. The sensation that our bladder is full is distinct from the need (urge) to empty it," and Nachev states, "It is surely right that urges must be dissociated from the sensations that in some cases prompt them." While it seems entirely reasonable to draw a conceptual distinction between an awareness of a bodily sensation, such as an itch, and the urge to scratch the itch (action), we are not entirely convinced that the phenomenology of common urges always supports this distinction. Rothwell & Edwards state "We can feel an itch or a 'desire' to scratch and we can decide to withhold the scratch; if the itch persists, then we may develop an urge to scratch," and he appears to equate awareness of a bodily sensation (itch) with the "desire" to execute an action (scratch the itch). We feel that this may more accurately reflect how we most often experience an urge. For instance, when we experience the sensation that our bladder is full, we also simultaneously experience the urge to empty our bladder; furthermore, it is not clear to us that these two sensations can be readily distinguished from one another.

However, as we noted in our article, it is not clear that we are ever aware of the bodily sensation that gives rise to the urge to yawn, and if we equate urges with the deferment or suppression of an action, then we also seem to be assuming that suppressing the urge for action differs from suppressing the bodily sensation giving rise to the urge. Clearly, future research is needed to understand how awareness of bodily sensation might differ from the urge for action.

ARE URGES-FOR-ACTION DEFINED BY INHIBITION?

In his commentary, Nachev makes a number of very important points with which we concur. Nachev states, "It is constitutive of the notion of an urge that it must precede the action it urges," and he proposes that "urges therefore inevitably involve inhibition of action." He also notes that it is "striking that the actions commonly associated with urges—both in the normal and in the pathological state—are usually of the kind that cannot be actively *chosen* but only *withheld.*" Finally, he argues that urges cannot therefore play a role in the execution of voluntary action as has been previously argued (e.g., Haggard, 2005).

Rivera-Rei et al. and Rothwell & Edwards in their commentaries also stress the importance of inhibition in defining an urge-for-action. Rothwell & Edwards argue, very reasonably, for a distinction between two interacting systems: a set of stimulus–response associations that may exist at all levels within the sensorimotor system, and a supervisory system that permits actions to be withheld or deferred. This in our view equates to the familiar distinction between "habit" and "intentional action" systems referred to in our article, and might be taken to indicate that the "inhibition" associated with the urge-for-action is explicit, goal-directed, and necessarily linked to the operation of a "supervisory" attentional system. If so, then we do not entirely agree with this proposal. Firstly, as recently pointed out by Aron (2007), the term *inhibition* is used very loosely within cognitive neuroscience, often without reference to any form of mechanistic account through which actions could be suppressed, withheld, or deferred. Aron distinguishes between psychological and physiological uses of the term *inhibition*, and suggests that at the psychological level "the notion of an active inhibitory process underlying cognitive control has been heavily challenged. Alternative explanations have been provided that explain cognitive control without recourse to inhibition as concept, mechanism, or theory" (p. 214). Even at the physiological level, where one encounters more mechanistic accounts, the nature of inhibitory mechanisms remains complex. For instance, an increase in thalamocortical activation can be brought about by the suppression of inhibitory neurons within the globus pallidus by striatal projections that are themselves inhibitory. In the current context, it may be again useful to consider the urge to urinate. In healthy adults, urination is assumed to be under "voluntary" control, whereas in young children, the elderly, and patients with neurological damage, urination can be involuntary and reflexive. Furthermore, the current state of this reflex system is dependent on sensory signals from the bladder and control signals from cortex, and voiding of the bladder is controlled by several mechanisms and necessarily involves the interaction of autonomic and somatic nervous systems. Specifically, during storage, the urethral sphincter muscles are tense and the detrusor muscle relaxed as a consequence of activation within the sympathetic nervous system. By contrast during voiding, activation within the parasympathetic nervous system causes the detrusor muscle to contract and the internal urethral sphincter to relax. Thus, the external urethral sphincter actually relaxes during micturition.

Here we take the view that the selection of action, including the inhibition of inappropriate responses, need not necessarily depend upon consciously directed supervisory mechanisms but can instead be based upon unconscious, effector-specific, control mechanisms. Evidence in support of this view can be found in a recent study reported by Sumner et al. (2007). Rivera-Rei and colleagues state, "It is hard to imagine a fully unconscious inhibition of the kinds of actions considered to be representative of urge-for-action (coughing, swallowing, yawning, etc.)." As it is not possible in this brief response to do justice to the evidence in support of this proposal, we urge interested readers to consult the article by Sumner et al. (2007).

ARE TICS INVOLUNTARY OR (UN)VOLUNTARY?

Several of the commentaries (e.g., Belluscio et al.; de Haan; Ganos & Hummel) referred to the fact that unwanted actions, such as the tics that occur in Tourette syndrome, or the compulsive behaviors associated with OCD, are not necessarily involuntary but should instead be considered (un)voluntary. This distinction is nicely elaborated upon within the Ganos & Hummel commentary, and we are very grateful to the commentators who raised this issue for clarifying this important distinction. We agree with the suggestion that further research is required to investigate similarities and differences between the urge-for-action associated with everyday behaviors and the urge-for-action associated with conditions such as Tourette syndrome and OCD.

ROLE OF THE INSULAR AND CINGULATE CORTICES IN EXPERIENCING URGES-FOR-ACTION

Both Rothwell & Edwards and Proverbio question the precise role(s) played by the insular and cingulate cortices in our experience of the urge-for-action. Such questions are entirely reasonable, and we freely admit that as yet we can only speculate on the roles played by these areas when we experience an urge-for-action in the context of everyday behaviors, or in the context of pathological conditions such as Tourette syndrome.

In our paper, we made use of quantitative meta-analytic techniques to investigate regions of overlap between brain-imaging studies of behaviors linked to the urge-for-action (e.g., swallowing, micturition, yawning). Interestingly, these analyses demonstrated significant overlap in only two regions: the right anterior insula and the mid-cingulate cortex, a finding that we interpreted as suggesting that these regions may have an important role to play more generally in representing the urge-for-action. Nevertheless, we are fully aware that both of these areas are frequently activated in human brain-imaging studies, across a wide range of behavioral domains, and that the correlational nature of brain-imaging studies makes it rather difficult to clarify the nature of the information-processing operations carried out within these regions.

In our view, the most important evidence that the insula may play a critical role in the experience of urges comes from lesion studies such as that reported by Naqvi, Rudrauf, Damasio, Rudrauf, and Bechara (2007), in which it was clearly demonstrated that smokers who had sustained damage involving the insula exhibited a disruption of their smoking addiction, described as "like their body forgetting the urge to smoke," that was not observed in individuals with comparable levels of brain damage that spared the insula.

In summary, our aim was to promote discussion of the nature and functional anatomy of the urge-for-action in both everyday behaviors and pathological conditions such as Tourette syndrome. We are grateful to the many colleagues who submitted commentaries on our paper, and encouraged by the many useful and insightful comments that we received. We feel certain that this topic will be the subject of much experimentation and debate in the near future and we look forward to participating in this process.

References from the Discussion Paper, the Commentaries, and the Reply

Ackermans, L., Duits, A., van der Linden, C., Tijssen, M., Schruers, K., Temel, Y., et al. (2011). Double-blind clinical trial of thalamic stimulation in patients with Tourette syndrome. *Brain, 134*, 832–844.

Albin, R. L., & Mink, J. W. (2006). Recent advances in Tourette syndrome research. *Trends in Neurosciences, 29*, 175–182.

Aron, A. R. (2007). The neural basis of inhibition in cognitive control. *Neuroscientist, 13*, 214–228.

Augustine, J. R. (1996). Circuitry and functional aspects of the insular lobe in primates including humans. *Brain research. Brain Research Reviews, 22*, 229–244.

Baier, B., & Karnath, H. O. (2008). Tight link between our sense of ownership and self-awareness of actions. *Stroke, 39*, 486–488.

Banaschewski, T., Woerner, W., & Rothenberger, A. (2003). Premonitory sensory phenomena and suppressibility of tics in Tourette syndrome: Developmental aspects in children and adolescents. *Developmental Medicine & Child Neurology, 45*, 700–703.

Bancaud, J., Talairach, J., Geier, S., Donis, A., Trottier, S., & Manrique, M. (1976). Behavioral manifestations induced by electric stimulation of the anterior cingulate gyrus in man. *Revue Neurologique, 132*, 705–724.

Behrens, T. E., Woolrich, M. W., Walton, M. E., & Rushworth, M. F. (2007). Learning the value of information in an uncertain world. *Nature Neuroscience, 10*, 1214–1221.

Blakemore, S., Wolpert, D., & Frith, C. (2002). Abnormalities in the awareness of action. *Trends in Cognitive Science, 6*, 237–242.

Bliss, J. (1980). Sensory experiences of Gilles de la Tourette syndrome. *Archives of General Psychiatry, 37*, 1343–1347.

Bohlhalter, S., Goldfine, A., Matteson, S., Garraux, G., Hanakawa, T., Kansaku, K., et al. (2006). Neural correlates of tic generation in Tourette syndrome: An event-related functional MRI study. *Brain, 129*, 2029–2037.

Burdakov, D., Luckman, S. M., & Verkhratsky, A. (2005). Glucose-sensing neurons of the hypothalamus. *Philosophical Transactions of the Royal Society of London, Series B, Biological Sciences, 360*(1464), 2227–2235.

Cameron, O. G. (2002). *Visceral sensory neuroscience: Interoception* (pp. 35–55). New York, NY: Oxford University Press.

Conelea, C. A., & Woods, D. W. (2008). Examining the impact of distraction on tic suppression in children and adolescents with Tourette syndrome. *Behaviour Research and Therapy, 46*, 1193–1200.

Craig, A. D. (2002). How do you feel? Interoception: The sense of the physiological condition of the body. *Nature Reviews Neuroscience, 3*, 655–666.

Craig, A. D. (2003). Pain mechanisms: Labeled lines versus convergence in central processing. *Annual Review of Neuroscience, 26*, 1–30.

Craig, A. D. (2009). How do you feel – now? The anterior insula and human awareness. *Nature Reviews Neuroscience, 10*, 59–70.

Damasio, A. (1999). The feeling of what happens: Body, emotion and the making of consciousness. London, UK: Heinemann.

Davenport, P. W. (2008). Urge-to-cough: What can it teach us about cough? *Lung, 186*, S107–S111.

Davenport, P. W., Sapienza, C. M., & Bolser, D. C. (2002). Psychophysical assessment of the urge-to-cough. *European Respiratory Review, 12*, 249–253.

Dayan, P., & Niv, Y. (2008). Reinforcement learning: The good, the bad and the ugly. *Current Opinion in Neurobiology, 18*, 185–196.

Deecke, L., & Kornhuber, H. H. (1978). An electrical sign of participation of the mesial "supplementary" motor cortex in human voluntary finger movement. *Brain Research, 159*, 473–476.

Desmurget, M., & Sirigu, A. (2009). A parietal-premotor network for movement intention and motor awareness. *Trends in Cognitive Science, 13*, 411–419.

Dehaene, S., Naccache, L., Cohen, L., Bihan, D. L., Mangin, J. F., Poline, J. B., et al. (2001). Cerebral mechanism of word masking and unconscious repetition priming. *Nature Neuroscience, 4*, 752–758.

Del Cul, A., Baillet, S., & Dehaene, S. (2007). Brain dynamics underlying the nonlinear threshold for access to consciousness. *PLoS Biology, 5*, e260.

Dijkerman, H. C., & De Haan, E. H. F. (2007). Somatosensory processes subserving perception and action. *Behavioral and Brain Sciences, 30*, 189–239.

Duggal, H. S., & Nizamie, S. H. (2002). Bereitschaftspotential in tic disorders: A preliminary observation. *Neurology India, 50*, 487–489.

Eickhoff, S. B., Amunts, K., Mohlberg, H., & Zilles, K. (2006). The human parietal operculum. II. Stereotaxic maps and correlation with functional imaging results. *Cerebral Cortex, 16*, 268–279.

Eickhoff, S. B., Laird, A. R., Grefkes, C., Wang, L. E., Zilles, K., & Fox, P. T. (2009). Coordinate-based activation likelihood estimation meta-analysis of neuroimaging data: A random-effects approach based on empirical estimates of spatial uncertainty. *Human Brain Mapping, 30*, 2907–2926.

Eimer, M., & Schlaghecken, F. (2002). Links between conscious awareness and response inhibition: Evidence from masked priming. *Psychonomic Bulletin & Review, 9*, 514–520.

Gallup, A. C., & Gallup, G. G., Jr. (2007). Yawning as a brain cooling mechanism: Nasal breathing and forehead cooling diminish the incidence of contagious yawning. *Evolutionary Psychology, 5*(1), 92–101.

Geangu, E., Benga, O., Stahl, D., & Striano, T. (2010). Contagious crying beyond the first days of life. *Infant Behavior and Development, 33*(3), 279–288.

Gerard, E., & Peterson, B. (2003). Developmental processes and brain imaging studies in Tourette syndrome. *Journal of Psychosomatic Research, 55*, 13–22.

Graybiel, A. M. (2008). Habits, rituals, and the evaluative brain. *Annual Review of Neuroscience, 31*, 359–387.

Haggard, P. (2005). Conscious intention and motor cognition. *Trends in Cognitive Sciences, 9*, 290–295.

Haggard, P., & Eimer, M. (1999). On the relation between brain potentials and the awareness of voluntary movements. *Experimental Brain Research*, *126*, 128–133.

Heise, K. F., Steven, B., Liuzzi, G., Thomalla, G., Jonas, M., Müller-Vahl, K., et al. (2010). Altered modulation of intracortical excitability during movement preparation in Gilles de la Tourette syndrome. *Brain*, *133*, 580–590.

Ibáñez, A., Gleichgerrcht, E., & Manes, F. (2010). Clinical effects of insular damage in humans. *Brain Structure and Function*, *214*, 397–410.

Jackson, S. R., Parkinson, A., Jung, J., Ryan, S. E., Morgan, P. S., Hollis, C., & Jackson, G. M. (2011). Compensatory neural reorganization in Tourette syndrome. *Current Biology*, *21*, 580–585.

Karnath, H. O., Baier, B., & Nagele, T. (2005). Awareness of the functioning of one's own limbs mediated by the insular cortex? *Journal of Neuroscience*, *25*, 7134–7138.

Karp, B. I., Porter, S., Toro, C., & Hallett, M. (1996). Simple motor tics may be preceded by a premotor potential. *Journal of Neurology, Neurosurgery, and Psychiatry*, *61*, 103–106.

Kühn, S., Haggard, P., & Brass, M. (2009). Intentional inhibition: How the "veto-area" exerts control. *Human Brain Mapping*, *30*, 2834–2843.

Kwak, C., Dat Vuong, K., & Jankovic, J. (2003). Premonitory sensory phenomenon in Tourette's syndrome. *Movement Disorders*, *18*, 1530–1533.

Lang, A. (1990). Patient perception of tics and other movement disorders. *Neurology*, *41*(2), 223–228.

Lang, S., Yu, T., Markl, A., Müller, F., & Kotchoubey, B. (2011). Hearing others' pain: Neural activity related to empathy. *Cognitive, Affective, and Behavioral Neuroscience*, *11*(3), 386–395.

Leckman, J. F. (2002). Tourette's syndrome. *Lancet*, *360*, 1577–1586.

Leckman, J., Walker, D., & Cohen, D. (1993). Premonitory urges in Tourette's syndrome. *American Journal of Psychiatry*, *150*, 98–102.

Libet, B., Gleason, C. A., Wright, E. W., & Pearl, D. K. (1983). Time of conscious intention to act in relation to onset of cerebral activity (readiness-potential). The unconscious initiation of a freely voluntary act. *Brain*, *106*, 623–642.

Lim, S. H., Dinner, D. S., Pillay, P. K., Luders, H., Morris, H. H., Klem, G., et al. (1994). Functional anatomy of the human supplementary sensorimotor area: Results of extraoperative electrical stimulation. *Electroencephalography and Clinical Neurophysiology*, *91*, 179–193.

Lowell, S. Y., Poletto, C. J., Knorr-Chung, B. R., Reynolds, R. C., Simonyan, K., & Ludlow, C. L. (2008). Sensory stimulation activates both motor and sensory components of the swallowing system. *NeuroImage*, *42*, 285–295.

Medford, N., & Critchley, H. D. (2010). Conjoint activity of anterior insular and anterior cingulate cortex: Awareness and response. *Brain Structure and Function*, *214*, 535–549.

Moore, J. W., Lagnado, D., Deal, D. C., & Haggard, P. (2009). Feelings of control: Contingency determines experience of action. *Cognition*, *110*, 279–283.

Moretto, G., Schwingenschuh, P., Katschnig, P., Bhatia, K. P., & Haggard, P. (2011). Delayed experience of volition in Gilles de la Tourette syndrome. *Journal of Neurology, Neuro- surgery, and Psychiatry.* Advance online publication. doi: 10.1136/jnnp.2010.221143.

Naqvi, N. H., & Bechara, A. (2008). The hidden island of addiction: The insula. *Trends in Neurosciences*, *32*, 56–67.

Naqvi, N. H., Rudrauf, D., Damasio, H., & Bechara, A. (2007). Damage to the insula disrupts addiction to cigarette smoking. *Science*, *315*, 531–534.

Obeso, J. A., Rothwell, J. C., & Marsden, C. D. (1981). Simple tics in Gilles de la Tourette's syndrome are not prefaced by a normal premovement EEG potential. *Journal of Neurology, Neurosurgery, and Psychiatry*, *44*, 735–738.

Olausson, H., Lamarre, Y., Backlund, H., Morin, C., Wallin, B. G., Starck, G., et al. (2002). Unmyelinated tactile afferents signal touch and project to insular cortex. *Nature Neuroscience*, *5*, 900–904.

Ostrowsky, K., Magnin, M., Ryvlin, P., Isnard, J., Guenot, M., & Mauguiere, F. (2002). Representation of pain and somatic sensation in the human insula: A study of responses to direct electrical cortical stimulation. *Cerebral Cortex*, *12*, 376–385.

Parkinson, A., Condon, L., & Jackson, S. R. (2010). Parietal cortex coding of limb posture: In search of the body schema. *Neuropsychologia*, *48*, 3228–3234.

Parkinson, A., Plukaard, S., Pears, S. L., Newport, R., Dijkerman, C., & Jackson, S. R. (2011). Modulation of somatosensory perception by motor intention. *Cognitive Neuroscience*, *2*, 47–56.

Pellijeff, A., Bonilha, L., Morgan, P. S., McKenzie, K., & Jackson, S. R. (2006). Parietal updating of limb posture: An event-related fMRI study. *Neuropsychologia*, *44*, 2685–2690.

Penfield, W., & Faulk, M. E. (1955). The insula: Further observations on its function. *Brain*, *78*, 445–470.

Picard, N., & Strick, P. L. (2001). Imaging the premotor areas. *Current Opinion in Neurobiology*, *11*, 663–672.

Platek, S. M., Mohamed, F. B., & Gallup, G. G., Jr. (2005). Contagious yawning and the brain. *Cognitive Brain Research*, *23*(2–3), 448–452.

Preuschoff, K., Quartz, R. S., & Bossaerts, P. (2008). Human insula activation reflects risk prediction errors as well as risk. *Journal of Neuroscience*, *28*, 2745–2752.

Proverbio, A. M., Adorni, R., Zani, A., & Trestianu, L. (2009). Sex differences in the brain response to affective scenes with or without humans. *Neuropsychologia*, *47*(12), 2374–2388.

Roessner, V., Wittfoth, M., Schmidt-Samoa, C., Rothenberger, A., Dechent, P., & Baudewig, J. (2011). Altered motor network recruitment during finger tapping in boys with Tourette syndrome. *Human Brain Mapping*. Advance online publication. doi: 10.1002/hbm.21240.

Rushworth, M. F. S., Mars, R. B., & Summerfield, C. (2009). General mechanisms for making decisions? *Current Opinion in Neurobiology*, *19*, 75–83.

Rushworth, M. F. S., Walton, M. E., Kennerley, S. W., & Bannerman, D. M. (2004). Action sets and decisions in the medial frontal cortex. *Trends in Cognitive Sciences*, *8*, 410–417.

Sander, K., Frome, Y., & Scheich, H., (2007). fMRI activations of amygdala, cingulate cortex, and auditory cortex by infant laughing and crying. *Human Brain Mapping, 28*(10), 1007–1022.

Schüermann, M., Hesse, M. D., Stephan, K. E., Saarela, M., Zilles, K., Hari, R., et al. (2005). Yearning to yawn: The neural basis of contagious yawning. *NeuroImage, 24*, 1260–1264.

Schultz, W., Dayan, P., & Montague, P. R. (1997). A neural substrate of prediction and reward. *Science, 275*, 1593–1599.

Serrien, D. J., Orth, M., Evans, A. H., Lees, A. J., & Brown, P. (2005). Motor inhibition in patients with Gilles de la Tourette syndrome: Functional activation patterns as revealed by EEG coherence. *Brain, 128*, 116–125.

Sirigu, A., Daprati, E., Ciancia, S., Giraux, P., Nighoghossian, N., Posada, A., & Haggard, P. (2004). Altered awareness of voluntary action after damage to the parietal cortex. *Nature Neuroscience, 7*, 80–84.

Sumner, P., Nachev, P., Morris, P., Peters, A. M., Jackson, S. R., Kennard, C., et al. (2007). Human medial frontal cortex mediates unconscious inhibition of voluntary action. *Neuron, 54*, 697–711.

Talairach, J., Bancaud, J., Geier, S., Bordas-Ferrer, M., Bonis, A., Szikla, G., et al. (1973). The cingulate gyrus and human behaviour. *Electroencephalography and Clinical Neurophysiology, 34*, 45–52.

Taylor, K. S., Seminowicz, D. A., & Davis, K. D. (2009). Two systems of resting state connectivity between the insula and cingulate cortex. *Human Brain Mapping, 30*, 2731–2745.

Thompson, S. B. (2011). Is yawning a warning, neurologically? *Webmed Central Neurology, 2*(2), WMC001610.

Tsakiris, M. (2010). My body in the brain: A neurocognitive model of body ownership. *Neuropsychologia, 48*, 703–712.

Tsakiris, M., Hesse, M. D., Boy, C., Haggard, P., & Fink, G. (2007). Neural signatures of body ownership: A sensory network for bodily self-consciousness. *Cerebral Cortex, 17*, 2235–2244.

Turkeltaub, P. E., Eden, G. F., Jones, K. M., & Zeffiro, T. A. (2002). Meta-analysis of the functional neuroanatomy of single-word reading: Method and validation. *NeuroImage, 16*, 765–780.

Turner, R. S., & Desmurget, M. (2010). Basal ganglia contributions to motor control: A vigorous tutor. *Current Opinion in Neurobiology, 20*, 704–716.

Vallbo, Å. B., Olausson, H., & Wessberg, J. (1999). Unmyelinated afferents constitute a second system coding tactile stimuli of the human hairy skin. *Journal of Neurophysiology, 81*, 2753–2763.

van Gaal, S., Ridderinkhof, K. R., van den Wildenberg, W. P., & Lamme, V. A. (2009). Dissociating consciousness from inhibitory control: Evidence for unconsciously triggered response inhibition in the stop-signal task. *Journal of Experimental Psychology: Human Perception and Performance, 35*, 1129–1139.

Wegner, D. M. (2002). *The illusion of conscious will.* Cambridge, MA: MIT Press.

Wolpert, D. M. (1997). Computational approaches to motor control. *Trends in Cognitive Science, 1*, 209–216.

Wolpert, D. M., & Ghahramani, Z. (2000). Computational principles of movement neuroscience. *Nature Reviews Neuroscience, 3*, 1212–1217.

Wolpert, D. M., Goodbody, S. J., & Husain, M. (1998). Maintaining internal representations: The role of the human superior parietal lobe. *Nature Neuroscience, 1*, 529–533.

Instructions for Authors

Please make sure your contact address information is clearly visible on the outside of all packages you are sending to Editors.

MANUSCRIPT SUBMISSION

Articles should be submitted through our ScholarOne Manuscripts online submission site: **http://mc.manuscriptcentral.com/pcns**

Cognitive Neuroscience publishes three kinds of article: (1) Short Reports (max. 4,000 words), (2) Discussion Papers (max. 8,000 words), and (3) Commentaries (max. 800 words). For more details about the style of each type of article, visit the Journal website at: **http://www.psypress.com/cognitiveneuroscience/submission**

FORMAT

Typescripts. The style and format of the typescripts should conform to the specifications given in the *Publication Manual of the American Psychological Association* (6th ed.). Typescripts should be **double spaced** on A4 paper, with adequate margins, and numbered throughout. The title page of an article should contain only:

(1) the title of the paper, the name(s), address(es), and e-mails of the author(s)
(2) a short title not exceeding 40 letters and spaces, which will be used for page headlines
(3) name and address of the author to whom correspondence and proofs should be sent, together with the telephone, fax, and e-mail numbers, as this helps speed of processing considerably.

Abstract. An abstract and 5 Keywords, should follow the title page on a separate page.

Headings. Indicate headings and subheadings for different sections of the paper clearly. Do not number headings.

Acknowledgements. These should be as brief as possible and typed on a separate page at the beginning of the text.

Permission to quote. Any direct quotation, regardless of length, must be accompanied by a reference citation that includes a page number. Any quote over six manuscript lines should have formal written permission to quote from the copyright owner. It is the author's responsibility to determine whether permission is required from the copyright owner and, if so, to obtain it.

Reference list. A full list of references quoted in the text should be given at the end of the paper in alphabetical order of authors' surnames (or chronologically for a group of references by the same authors), commencing as a new page, typed double spaced. **Titles of journals and books should be given in full**, e.g.:

> ***Books:***
> Baddeley, A. D. (1999). *Essentials of human memory.* Hove, UK: Psychology Press.
>
> ***Chapter in edited book:***
> Plomin, R., & Dale, P. S. (2000). Genetics and early language development: A UK study of twins. In D. V. M. Bishop & L. B. Leonard (Eds.), *Speech and language impairments in children: Causes, characteristics, intervention and outcome* (pp. 35–51). Hove, UK: Psychology Press.
>
> ***Journal article:***
> Schwartz, M. F., & Hodgson, C. (2002). A new multiword naming deficit: Evidence and interpretation. *Cognitive Neuropsychology, 19*, 263–288. doi:10.1080/02643290143000187

Color figures will be included at no cost to the author. Figures should only be used when essential and the same data should not be presented both as a figure and in a table. Each of these should be submitted as a separate file (preferably tif, jpg, or eps) and saved at a high resolution (at least 300 dpi).

(Continued from back cover)

For Product Safety Concerns and Information please contact our EU representative GPSR@taylorandfrancis.com Taylor & Francis Verlag GmbH, Kaufingerstraße 24, 80331 München, Germany

Batch number: 10733530

Printed by Printforce, the Netherlands